GT3
The Unrealised Dream
The Story of Britain's Last Gas Turbine Locomotive

GT3
The Unrealised Dream
The Story of Britain's Last Gas Turbine Locomotive

TIM HILLIER-GRAVES

PEN & SWORD TRANSPORT

AN IMPRINT OF PEN & SWORD BOOKS LTD.
YORKSHIRE – PHILADELPHIA

First published in Great Britain in 2025 by
Pen and Sword Transport
An imprint of
Pen & Sword Books Ltd.
Yorkshire - Philadelphia

Copyright © Tim Hillier-Graves, 2025

ISBN 978 1 03610 666 9

The right of Tim Hillier-Graves to be identified as author of this work has been asserted by him in accordance with the Copyright, Designs and Patents Act 1988.

A CIP catalogue record for this book is available from the British Library.

All rights reserved. No part of this book may be reproduced or transmitted in any form or by any means, electronic or mechanical including photocopying, recording or by any information storage and retrieval system, without permission from the Publisher in writing.

Typeset in Palatino by SJmagic DESIGN SERVICES, India.
Printed and bound by Printworks Global Ltd, London/Hong Kong.

Pen & Sword Books Ltd. incorporates the imprints of Pen & Sword Books: After the Battle, Archaeology, Atlas, Aviation, Battleground, Discovery, Family History, History, Maritime, Military, Politics, Select, Transport, True Crime, Fiction, Frontline Books, Leo Cooper, Praetorian Press, Seaforth Publishing, Wharncliffe and White Owl.

For a complete list of Pen & Sword titles please contact

PEN & SWORD BOOKS LIMITED
George House, Beevor Street, Off Pontefract Road, Hoyle Mill, Barnsley, South Yorkshire, England, S71 1HN.
E-mail: enquiries@pen-and-sword.co.uk
Website: www.pen-and-sword.co.uk

or

PEN AND SWORD BOOKS
1950 Lawrence Rd, Havertown, PA 19083, USA
E-mail: uspen-and-sword@casematepublishers.com
website: www.penandswordbooks.com

CONTENTS

	Acknowledgements	6
Prologue	Marylebone Goods Yard May 1961	8
Chapter 1	An Age of Opportunity and Change	16
Chapter 2	A Slow Burning Revolution	29
	Colour Section – GT3 – A Design History in Colour	65
Chapter 3	A Slow Burning Ambition	81
Chapter 4	Out of Time	112
Epilogue	GT3 – The Unrealised Dream	137
Appendix 1	Engineering and Aesthetics – How GT3 Took Shape	145
Appendix 2	GT3 Under Test at Rugby	155
Appendix 3	GT3 Dynamometer Car Tests	168
Appendix 4	John Hughes's Project Summary as Prepared in April 1962	172
	References and Sources	176
	Photographic Sources/Credits	177
	Index	178

ACKNOWLEDGEMENTS

It is most unlikely that this book would have been written without the help of material carefully collected by John Hughes, GT3's creator. Having researched many railway related subjects over a long period of time, I can say without fear or favour that the history he recorded about a single, complex design project, has few equals. Following John's death in 1977, his wife Felicity and family preserved this legacy and donated some of this collection to the National Railway Museum where it may now be consulted in Search Engine.

Sadly, I didn't meet John, but did meet his wife who allowed me to copy his papers and gave me permission to use the material as I saw fit, hoping that one day her husband's efforts might finally be recognised. I hope that this charming lady, who died in 2009, would think this book is an appropriate memorial to his work.

I was also helped in this endeavour by research my late uncle, Ronald Hillier, undertook and items he collected. Then there are my three old friends, David Neal, Jack Constantine of New Jersey and Phil Atkins, late of the NRM, whose help and advice was always given so generously.

In Britain we are lucky to have many institutions that have meticulously preserved as much material as possible for all to see and research. But sadly, much was destroyed by British Railways in the 1960s during a regrettable bout of 'good housekeeping'. However, more by luck than judgement, a great deal was saved by concerned individuals and this now exists in public and private collections. There are significant

GT3, and its tender (seen on the far right of the engine), undergoes final assembly at the Vulcan Foundry Works in late 1960, as captured by its designer in chief, John Hughes. Luckily, he kept a detailed record of the entire gas turbine project that he and English Electric were committed to from the late 1940s to the early 1960s. In doing so, he provided historians with a unique and invaluable record of a fascinating, but ultimately doomed, venture. (JH)

gaps and where these exist, many people have committed time and energy to seek out information and publish the results of their work wherever possible. And by this diverse means an important part of our social and engineering history has been preserved to illuminate our understanding of the past.

In writing this book, I drew heavily on three institutions in particular. First and foremost, there is the National Railway Museum and its extensive archives, then 'Discovery' at The National Archives and, finally the library of the Institution of Mechanical Engineers in London. The staff who manage these collections do so with great skill and always proved friendly and co-operative when answering my many questions, despite their busy schedules.

To all the people who helped me to research and write this book, I give my thanks and hope I have done justice to all that they have contributed. Ultimately though, all an historian can do is to sift and consider all material and reach a judgement that he or she thinks honestly reflects past events. There will, undoubtedly, be alternative views or conclusions, but that is as it should be. I don't think there's ever a final word and new material may be found to allow fresh interpretations to be made.

Prologue

MARYLEBONE GOODS YARD MAY 1961

For fifty years, the Institution of Locomotive Engineers had provided a professional focus for its members.

It had become a centre of excellence, where information was gathered and shared with people from across the world. By slow degrees it had grown in stature and accomplishment, pursuing its primary aim of recording and conveying thoughts, ideas and discoveries. It was to locomotive engineers what the BMA and the *Lancet* are to the medical profession and RIBA to architects. But by 1961 many felt that the Institution's greatest days were over. With the passing of steam locomotion, the slow collapse of the railway industry in Britain and the growth of manufacturing overseas, its central focus seemed to be drifting away. Yet even allowing for this

The sight greeting the many hundreds of people attending the ILocoE's railway exhibition to commemorate its 50th Anniversary, including the author and his uncle. GT3 was positioned in the left hand queue of locomotives behind two steam engines and a London Midland BR Doncaster designed and built electric AL5 (later Class 85). (THG)

gradual erosion of the UK's railway engineering base, the work of the Institution was still seen as essential by many.

At Marylebone in 1961, its fifty years of life were celebrated with an exhibition of engines and rolling stock, all deemed to be the cream of Britain's railway history, and to reflect upon what might lie ahead. But by the end of the decade, with membership and funds dwindling, the organisation was wound up, its last independent strands absorbed by the Institution of Mechanical Engineers as a railway sub-division. With hindsight, the May 1961 exhibition could be seen as a last hurrah for this professional body; an end and a beginning with others taking up the reins in championing the cause of this distinctive area of science.

The 50th Celebration was spread over three days. Wednesday, 10 May saw a Jubilee lunch at the Dorchester Hotel in Park Lane, attended by the highly divisive figure of Ernest Marples, the Minister of Transport. This was followed by a lecture entitled 'The Locomotive of the Future', delivered by Sir Brian Robertson, Chairman of the British Transport Commission at the IMechE's Headquarters in Great George Street. Early the next morning, a special train was commissioned to run from Paddington to Bath, pulled by a Class 42 Warship diesel. After lunch, at the Limpley Stoke Hotel, the afternoon was spent at Longleat House before an evening return to Paddington from Westbury. But despite all these celebrations, for many the exhibition on 12 May was the most important and the most memorable part of the commemoration.

The Goods Station at Marylebone had been taken over for the exhibition a few days earlier, with a number of carefully selected exhibits gradually being brought

A publicity photo taken shortly before GT3 was placed on display at Marylebone. From early in its life, the English Electric PR team had tried very hard to get stories into the press about this new locomotive and circulated a great deal of information to those of importance in the railway industry. If it failed to gain favour it would not be the want of trying. (THG)

In a highly burnished state, GT3 and other engines are lined up for display on 10 May 1961, all having been cold hauled to the site a little earlier. (JH)

to together, cleaned and prepared for inspection. The Institution's Council had chosen items which they thought best summed up the development of locomotives over the previous fifty years, but were naturally restricted by availability. All guests and ticket holders passed through a decorated archway into the carefully laid out display arena and the weather held for the day and into the 13th when the exhibition was due to close.

Most of those invited gathered by 9.45 to be in place for the arrival of the Duke of Edinburgh and his entourage. Each exhibit was covered by an expert in a given field to guide the Duke and answer any questions he might have. But with most guests being Institution members there was no shortage of specialists on hand to explain the complexities of the engines and rolling stock laid before him. Even such famous railway veterans as Sir William Stanier, Oliver Bulleid, Bert Spencer, Robert Riddles, Roland Bond and Ernest Cox were there.

For three hours the Duke toured the exhibition, observing and discussing each locomotive, before departing in the cab of a Warship diesel for Windsor. And once the niceties of this official visit were over, members and public alike could clamber over the exhibits. Gresley's A4 *Mallard*, BR's 71000 *Duke of Gloucester* and 92220 *Evening Star* and Midland Compound number 1000 represented the old order, whilst the future was symbolised by the latest in diesel and electric traction. But there was one lone addition that fitted into neither camp – an experimental oil-fired gas turbine engine, painted a distinctive brown, with mechanical transmission, built by English Electric and designated GT3.

If anyone believed that this striking and unusual engine might represent an alternative future, they were echoing the thoughts of its chief designer who had doggedly pursued the concept for nearly a decade and a half. John Hughes was a design engineer who strongly believed in the principle of locomotives being powered by turbines. And in this he found a ready supporter in Stanier, who, during the 1930s and early '40s, had also toyed with this technology with some success.

Once the Duke of Edinburgh had left, Stanier sought out the younger man and spent a good part of the afternoon talking to Hughes about this work. Sadly, by then the one turbine engine the London Midland and Scottish Railway (LMS) had built was long gone – converted to conventional form in 1952 and then written off following the disaster at Harrow a few months later. Hughes had known it well, often seeing it passing by when working for Rolls Royce as a junior designer, and on two occasions had ridden behind her. He even had a set of drawings, given to him by Tom Colman, Stanier's Chief Draughtsman, many years earlier. As the years passed, he had drawn inspiration from this work and kept referring back to it as his own attempts to develop the science met with difficulties. The LMS's most famous CME was clearly moved by this, as John Hughes later recalled, and for a time they corresponded and met at Institution functions:

Sir William was an inspiration to many and never let an occasion pass when he didn't offer words of advice or support, even though advancing in years and slowing up considerably. Even when very old he possessed a dynamism that made me appreciate just how he had managed to achieve so much in his career. He had inner steel, wide scientific knowledge, considerable skill and a strong sense of humanity.

He talked at length about Turbomotive and the problems he had faced in bringing the locomotive into service and then getting the best out of her, in the face of opposition and lack of appropriate support. He was still of the opinion that the running department had failed to appreciate the engine's need for special handling and had let the project down. He was also convinced that many more should have been built, principally Princess Coronations and 8Fs. Work by Ljungstrom's in Sweden had proved the value of steam turbines, with their enhanced performance and less maintenance, and 6202 had begun to demonstrate the value of the technology on high speed services. On one occasion he described the lack of support he received in pushing ahead with a more ambitious programme as 'short sighted' and a missed 'opportunity'.

It seemed to me that Sir William saw GT3 as a natural successor to his work and powered by a fuel more suited to a turbine engine than coal. However, he did think electric transmission held some advantages over our mechanical version.

After many setbacks and trials, it seemed we had

Prince Philip, the ILocoE's guest of honour, questions John Hughes (far left) about the gas turbine engine he had designed and then project managed during construction. (RH)

Above and right:
William Stanier was a welcome guest at Marylebone (above left) and, apparently, showed great interest in John Hughes' work. It was, after all, a worthy successor to Stanier's 1930s built steam powered 4-6-2 *Turbomotive*, though neither could compare to the magisterial Coronation Class, one of which (the soon to be scrapped 46240 *City of Coventry*) is pictured here with GT3 in the distance. (RH)

produced an engine that was the beginning of a new era. All we needed to do, in 1961, was gain acceptance and the approval of the powers that be to continue. But during 1962 all our efforts came to nought when the programme was cancelled and GT3 was later broken up.

So the Marylebone Exhibition came to represent the high point of Hughes' achievement and also a farewell to him and his engine. Yet another scientific 'what if' that came to nought.

Science and discovery go hand in hand. One without the other is unthinkable. Enquiring, creative, analytical minds are drawn to the challenge of research and innovation. If it were otherwise humans wouldn't have advanced far beyond caves and a language consisting of grunts. Fame and profit are also a spur to these acts, but the truly great scientists tend to be driven by more than the acquisition of glitter or prizes. To those with integrity the act of creation is all that matters. It may not always end in a successful outcome but can still add something to the sum total of human knowledge and spark other advances of sometimes greater significance.

Those who succeed are probably reaping the benefit of the collective knowledge of centuries, but, equally so, owe much to luck as well; being in the right place at the right time, where chance, opportunity and need come together. But this tends to be seen only in retrospect, when success has been achieved and fame assured. Yet when the challenge was still fresh in the mind, the problems presented must have seemed insurmountable, with

GT3 with English Electric engineers and BR crew ready to begin trials with the completed engine. Hughes is in the group examining the new tender. (JH)

success a distant, unobtainable goal. And it was here that the skills of great scientists are revealed; a determination to overcome obstacles, to commit themselves even though their work may be 5 per cent inspiration and 95 per cent perspiration and pursue, relentlessly, a goal, even when that target seems unattainable.

Science covers many fields of human endeavour, yet central to all that is achieved is the engineer, a title that hasn't changed over the centuries, but encompasses so many different, often diverse skills. Hughes, though ultimately unsuccessful in this endeavour, demonstrated all that was best about his profession and his creed.

When the future of GT3 still seemed assured, sanctioned, or so it appeared, by the blessing of being an exhibit at such a prestigious event, John Hughes had been invited to present a paper to the ILocoE on his 'prototype gas turbine locomotive'. On 14 December 1961, he gave his first talk to members, which was later repeated in London and Leeds, and set out the advantages of such an engine. 'It gives high power with compactness and light weight, together with excellent reliability… there is an appreciable gain in

There are not many photographs of GT3 pulling trains during 1961 when undergoing rigorous tests on the mainline, but what there are show, to my mind, that this modern engine, with its dated steam locomotive 4-6-0 based outline, did look the part. (RH)

tractive effort at speed…. it requires very little maintenance and gives high availability… it offers the possibility of burning lower grade fuels.'

These elements were the Holy Grail of locomotive design and, if achieved, would have made GT3, and all that followed, worthy acquisitions for railway companies across the world. Yet within months the project had been cancelled and the locomotive consigned to the scrapyard, BR having decided that it didn't fit into its diesel or electrification programmes. Stripped of this support, and its main customer, English Electric could do little more than run down its gas turbine group and focus on delivering more traditional engines. After that Hughes seems to have become disenchanted with his employer and so left for pastures new and then, in turn, started his own company.

So the development of the mechanical transmission gas turbine locomotive in Britain came to an end. Another worthy idea that did not find favour or simply couldn't improve on other forms of traction, except steam, which was dying anyway. Work on gas turbine-electric locomotives continued for a while and had one notable success in the USA and two working examples in the UK. But even these didn't lead to greater acceptance or production. BR's prototype Advanced Passenger Train was turbine powered, but this was also dropped in favour of an electric system, when British Leyland, the manufacturers ceased production of its turbine. Yet, despite these setbacks, Hughes never lost his belief in the concept and advocated it to the end of his life.

Today we can look back and applaud these brave attempts to seek better ways of powering transport systems, especially if they offer solutions better suited to our environmental needs. A time may come when they might be revived, but for the moment this research remains an historical idiosyncrasy to be admired and debated and placed inevitably in the 'what if'

category. But in the Forties and Fifties, when the concept was new and exciting, it did transfix many clever engineers and offered the hope of great advances. This book seeks to capture the excitement of this work, with GT3 and John Hughes at its core.

I was lucky to have been taken to the Marylebone Exhibition by my late uncle and seen GT3 for myself. I have always been drawn to unusual and novel ideas and this engine, resplendent in her non-standard livery, ticked many boxes for me. I entered her cab to enjoy the sensation that only boyhood seemed able to capture and looked down at a sea of faces as intrigued by the sight of this unique engine as I was. Being only ten, I couldn't even begin to appreciate the technology, but enjoyed the experience none the less, my only disappointment being the side windows, which were reminiscent of those used on portacabins or prefabs. Were they an afterthought sponsored by economy, or had they been purloined from a building site next to the English Electric factory?

Sitting alongside new electric and diesel locomotives, with beautiful but archaic steam engines gathered nearby, GT3 presented a slightly strange picture. It was almost as if she had been dropped by aliens hoping to slip her into this esteemed company. With a steam loco wheelbase and the angularity of a diesel body, she presented an ambiguous picture. And I suppose this is why she took my eye long enough to deflect me from my inevitable path to *Mallard*, sitting only yards away.

As I look back at John Hughes' remarkable experiment I marvel at his tenacity and skill in taking an idea and pursuing it, no matter what obstacles were placed in his way. Not for him acquiesence or acceptance of the inevitable, but a dogged determination to pursue a concept to a workable conclusion. In so doing, he came within an ace of success, but was eventually beaten by BR's drive towards a diesel and electric future. With the benefit of hindsight it is easy to see that Hughes' efforts were doomed from the beginning and went on far longer than necessary to prove a point – it worked, but there were, as BR's managers believed, better options to pursue.

In truth this is a 'what might have been story' which, although ultimately unsuccessful, tells us much about the nature of design engineering and the need for determined, often radical, thinkers who are prepared to push back scientific boundaries to discover new, perhaps even better, solutions. In this case, GT3 came close to being the first of a type of engine that might have changed the direction Britain's railways took as steam came to an end. As such it is a near miss worthy of remembrance.

What might have been – a successful gas turbine prototype that British Rail could have developed in greater numbers and so moved their direction of travel in a better direction, or so some believed. GT3 caught at speed in 1961 pulling a passenger service. (JH)

Chapter 1
AN AGE OF OPPORTUNITY AND CHANGE

Even when nearing the peak of their development in the 1930s, steam locomotives were still recognisably based on the model George Stephenson predicated nearly a hundred years before. It was a form of locomotion that had served the country well and become stronger, faster and more reliable in the process. But the concept was constrained by its own technical limitations and the labour-intensive nature of their operation. In a country grown rich on coal and cheap labour, its continuing life wasn't surprising, but the rapid development of the internal combustion engine, the availability of cheap oil and the gradual expansion of the national power grid presented other possibilities to be explored. So locomotive designers began to look beyond coal fired technology to more fuel-efficient solutions.

There were also huge social drivers for change in play. The post-Second World War years saw a kindling of expectation that lives would improve significantly after decades of austerity and sacrifice. Wealth in Britain had always been held by a tiny, very privileged minority and a fairer share for all had become a national chant in righting the wrongs of centuries. This inevitably coalesced around the need for better housing, better schools, modern health and welfare services, improved pay and working conditions and an up to date, cleaner transport system. The demands were huge, but sadly not matched by the parlous state of Britain's coffers, which had been stripped bare by two wars. So the process of change had to be a gradual one, with advances measured against the near bankrupt state of Britain's economy and its continuing and costly overseas commitments.

Nevertheless, the die had been cast and it would have been a foolish politician who ignored the increasing clamour for change or tried to inflict a 'belt tightening' philosophy again on a public grown weary of such a refrain. And as the Fifties gave way to the Sixties, the first shoots of recovery were becoming only too apparent and change had developed a seemingly unstoppable critical mass – even on the railways, where steam still dominated and safety was often compromised by poor leadership and lack of cash.

It was hoped that nationalisation of the railways in 1948 would lead to much needed modernisation, but the nation's long reliance on coal was hard to break. With the mining industry employing many hundreds of thousands of people in deprived areas and the country relying on its output to reduce the national debt, its continued production and use was inevitable and unavoidable.

Like the captain on the bridge of his ship John Hughes looks down from his creation, by then with its chasis complete and power plant installed but with no cab or body fitted. (JH)

So it surprised few when BR's leaders and their political masters chose to build a large new fleet of steam locomotives, rather than invest heavily in other technologies. It was so different in the USA, France and even Germany, where Allied bombers had reduced its cities, towns and infrastructure to so much rubble. Here steam was allowed to continue where necessary, but diesel and electric locomotives were soon taking over.

In each country, as in Britain, the 1930s had seen some far-seeing designers begin the process of change, only to see the coming of war put paid to this work for a time. However, some important advances had been made by then. In Britain, the LMS had produced diesel shunters and a streamlined diesel multiple unit, while the

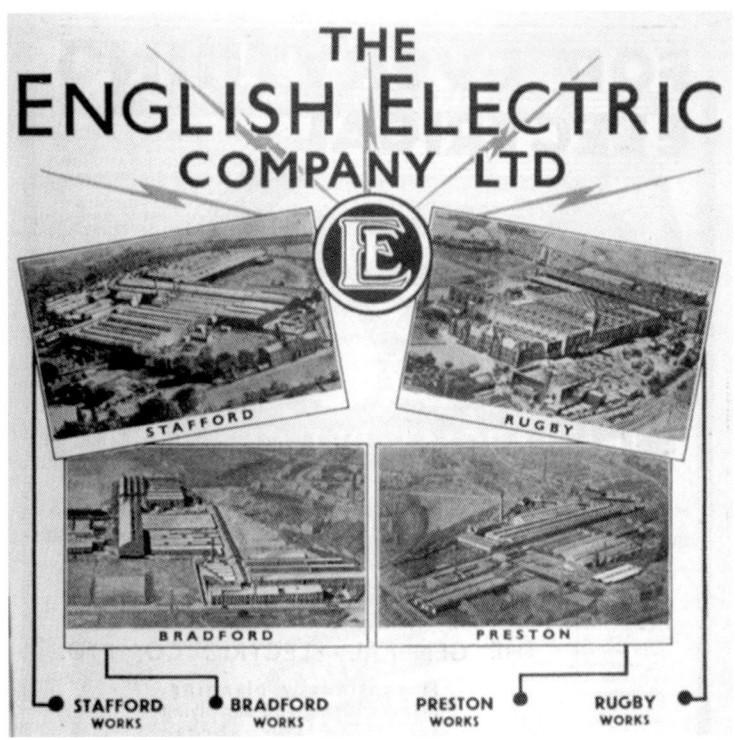

Left and below: **Having been** formed in 1918, English Electric began specialising in the production of large capacity transformers and turbines. During the Second World War the company expanded considerably developing armaments – tanks and aircraft in particular – for the War Department (above right). To help fill the void created by the end of the conflict they turned their attention to manufacturing domestic products and more. By the 1960s the company had expanded its portfolio to include nuclear reactors, guided missiles and military aircraft. In addition, locomotives had become a core and seemingly very profitable area of work which they sought to develop whenever they could. The creation of GT3 seemed to be a happy combination of their acquired knowledge of turbines and locomotives, so opening up a new field to be exploited. (Author)

A scene typical of Britain's railways in the post Second World War years – a tired network ripe for modernisation and still reliant on steam to keep it going. For the enthusiast a wonderful sight, for the economists and day to day users a system not fit for purpose. Here a rather shabby ex-LMS 5MT (possibly 45449 which was built by Armstrong Whitworth in 1937 and remained in service until virtually the end of steam in 1968) pulls a good-sized load at an unrecorded location. (Author)

London North Eastern Railway (LNER) had actively pursued a policy of mainline and commuter line electrification. Meanwhile, the Southern Railway already had a well-established electrification programme which they planned to extend across the network as soon as they could, soon adding experiments with diesel-electric locomotives as a second string to their bow. Meanwhile, the Great Western Railway (GWR) had an active diesel railcar programme and were beginning to toy with turbine designs, soon looking abroad where successful experiments were being conducted with diesel-hydraulic powered engines.

Sadly, when their time came, BR seems to have placed little emphasis on these programmes and allowed steam to dominate their thinking until well into the 1950s. Not so some of Britain's major industrial

It was hoped that nationalisation would lead to much-needed modernisation programmes for locomotives and rolling stock. BRs response, faced with lack of money and, perhaps, a will to change, was to build nearly 1,000 new standard locomotives and allow many existing, inherited programmes to continue. With hindsight this would be regarded as a missed opportunity that set back true modernisation by ten years or more. (Above) Here the first of 55 new 7P 4-6-2 Class Pacifics (No. 70000 *Britannia*) rolls of the production line at Crewe in 1951. In due course one 8P and ten 6Ps would be added to the fleet. (Author)

concerns, who saw the future more clearly and were prepared to invest and experiment on 'speculate to accumulate' ventures. Having grown large on wartime contracts for war materiel and machinery and having attracted the best scientists and engineers in the process, they were now well placed to take forward speculative ideas and forge a new commercial path. And of these English Electric were, perhaps, the best known and probably the most ambitious.

This huge conglomerate came into existence during 1918 with the merger of engineering companies such as Coventry Ordnance, Phoenix Dynamo and Dick, Kerr and Co. By 1945 this new company had expanded its business into such things as heavy electrical and mechanical plant, domestic appliances, steam turbines, aviation and railway projects at home and overseas. The last of these included, electric locomotives and multiple units for New Zealand Railways, electric motor cars for Warsaw's transport system, diesel multiple units for the Egyptian State Railway and electric shunting locomotives for the National Coal Board.

After the Second World War this expansion continued and sensing that railway companies around the world might soon be ditching steam for diesel and electric locomotion, English Electric began investing heavily in this developing technology. New design divisions were set up to explore these ideas, while at the same time the company put together a strong marketing team to help sell these new products to potential customers. If Britain's economy had been more buoyant in these post

Above and left: **In the mid-1950s,** with BR finally setting their sights on a true diesel and electric based modernisation programme, English Electric developed their experimental DP1 Deltic Co Co locomotive for evaluation by potential customers. In its striking blue livery, it attracted great interest and BR began a prolonged testing period with the engine, eventually resulting in an order for twenty-two. However, not all, as the cartoon above reveals, were in thrall to a diesel world and regretted the gradual erosion, then disappearance of steam locomotion. (Author)

war years, and British Railway's managers less wedded to coal and steam, this effort might have borne fruit much earlier than it did. Nevertheless, they pressed on with their work hoping that they, and other companies, might encourage BR to change their procurement strategy. This they did in 1954/55 when publishing an updated Modernisation Plan with diesel and electric locomotion at its core.

In response to this, English Electric introduced their prototype DP1 Co Co Deltic diesel during 1955 for BR to evaluate. After prolonged periods of testing, with both the London Midland and Eastern Regions, twenty-two were purchased and entered service on BR's East Coast Mainline in 1961/62. But at the same time they had begun successfully developing other diesels, most notably the Type 1 Bo Bo Class that began appearing in 1957 and the Type 4 Co Cos that began to enter service the following year. With the possibility of even more contracts, English Electric were soon pursuing other options as BR began to shrug off the age of steam and slowly advance. It was at this stage that the concept of turbine power began entering their development plans and soon they began recruiting design engineers, with experience in this field, to further these ideas.

The two world wars had many negative effects on the societies caught up in their traumas, but there were some positive developments which cannot be ignored. In war, scientists have to work hard to produce new and better weapons to secure an advantage over an enemy or simply to counteract a telling advance made by the other side. So the speed in which science moves forward can be dazzling with many applications conceived that have longer term benefits that enrich and benefit a peacetime world. Rapid advances in medicine necessary to treat the millions of wounded are an easily perceived outcome, but other sciences also benefitted from this work, most notably in the development of aircraft. Here

During the 1950s, English Electric became very closely linked with BRs attempts to modernise its locomotive fleet and played a central role in this programme. Sometimes they were a supplier of parts to BR, who produced the engines themselves, and on other occasions they were contracted to build the entire locomotive. Either way, the company's marketing team promoted their work as this widely circulated advertisement makes clear. As a result English Electric became one of BR's primary suppliers. (Author)

the search for greater thrust from engines and lighter, stronger materials – in the never-ending search for the most effective power to weight ratio – had outcomes relevant to other industries, most notably in automotive, ship and locomotive design. Then there was the issue of aerodynamics to consider and the ways in which forward resistance might be reduced to allow higher speeds and lower fuel consumption.

All this led to a greater concentration on the science of metallurgy and the way their chemical, physical, atomic properties might be shaped to form alloys that might be stronger than their constituent parts. Nigel Gresley identified the benefits the railways might derive from this discipline during the 1920s and '30s and quickly sought to apply its lessons wherever he could. At the same time, he allied these advances in material to a study of shape, using wind tunnels as a means of establishing the most effective aerodynamic form. Here he quickly realised that air-smoothing could reduce atmospheric resistance considerably and so pave the way for faster, more economic rail vehicles. He also studied the way engines might be more effectively balanced to improve their riding qualities, employing such specialists as William Dalby to do so. As a result of all this, he produced his record breaking A4 Pacifics in 1937 and, but for the war, would probably have taken these experiments further and applied them to LNER'S growing electrification programme.

The other process that increasingly influenced locomotive design and day to day running in the post war years was an advance in precision engineering. This grew in importance with the development of optical measuring equipment and techniques in Germany during the 1930s which made finer engineering tolerances more easily achieved. Britain was slow to develop a domestic version as Kenneth Cook, who became BR's Mechanical and Electrical Engineer at Swindon in January 1950 and chief advocate of this process, related in a paper to the Institute of Locomotive Engineers:

> At the Machine Tool Exhibition in 1952 a British optical exhibit was noticed that appeared capable of development although at that time it had no reference to locomotives. The makers became very anxious to co-operate and quite quickly a method, utilising instead of a collimator, a reflecting mirror fixed parallel to and in line with a straight edge method proved much simpler than the German (version), and capable of proceeding very much further in the quest for accuracy. It became known as the Auto-Reflexion.

This apparatus proved fully usable in the construction of locomotives to ensure initial accuracy and during general repairs when it can ensure and maintain accuracy over a long working life.

The transition from steam to diesel and electric motive power is almost complete, as witnessed by this Brush Type 2 A1A-A1A (No. D5642) locomotive. The engine is idling between turns at King's Cross, now no longer the preserve of Gresley's Pacifics, all of which have now been withdrawn with most quickly reduced to scrap metal. (Author)

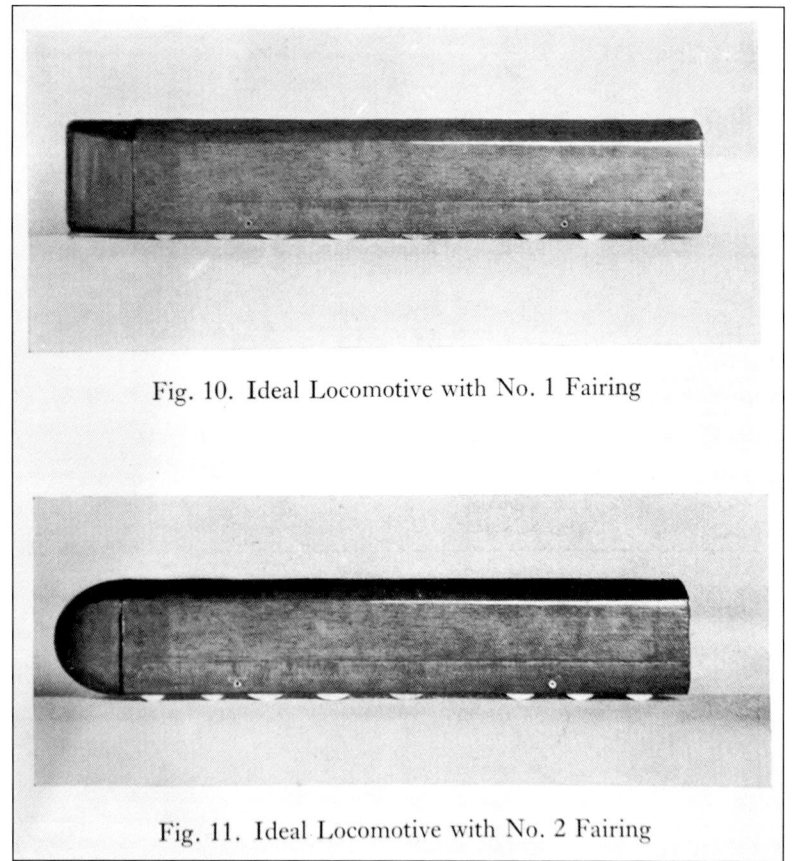

Fig. 10. Ideal Locomotive with No. 1 Fairing

Fig. 11. Ideal Locomotive with No. 2 Fairing

During the 1930s, new or evolving sciences began to impact on locomotive design, construction and in service maintenance needs. This was none more so than in the application of evolving aerodynamic principles and wind tunnel testing. Aviation had highlighted the benefits to be gained from this area of research and the railways began to apply some of the lessons learnt when designing locomotives for high-speed duties (above left shows the results of experiments to show how the shape of trains might be modified to reduce forward and sideways resistance). The development of optical measuring equipment by Taylor, Taylor and Hobsons (above right) in Britain, following on from similar work in Germany in the 1930s, revolutionised the process of constructing and maintaining aeroplanes but was equally well applied to railway locomotives. The Eastern Region, under Kenneth Cook, applied these techniques most notably to their Pacifics making good engines even better. The design and construction of diesel, electric and turbine locomotives would see these principles applied even more fully and effectively. (Author)

After successfully applying these new techniques at Swindon, then as Mechanical and Electrical Engineer at Doncaster, he was convinced of the advantages to be gained. With this in mind, he proffered his considered view of the through life benefits accruing and their beneficial effect when applied to newly constructed locomotives:

We should probably all agree that the economical criterion of locomotive performance is cost per mile in similar conditions of operation and one of the greatest factors in producing low cost per mileage obtained between heavy repairs. In carrying out a heavy repair, the dismantling and erecting costs are fairly constant whatever the mileage and a higher mileage enables these to be spread and to produce a lower overall figure. High precision in basic details of a locomotive can make a big contribution to economy.

With steam being replaced by diesel and electric locomotives, which were much more advanced in design, the need for greater accuracy when constructing and maintaining them was only too obvious. So more precise forms of optical measurement gradually entered into railway workshops across Britain with engineers learning and applying these new skills to sit alongside the long and well established competences so necessary for locomotive production and servicing. However, there is one other important consideration in the process of creating new designs – an effective, carefully monitored, well calibrated testing regime.

From early in his career with the LNER, Gresley had been very aware of this need, as indeed had George Churchward at Swindon earlier in the century. There had always been some sort of 'on the road' testing, usually carried out by engineers, balanced precariously behind a crude wooden shelter built around an engine's smokebox. Although better than nothing, the information this generated was limited, but then the railway companies developed dynamometer cars that began to reveal much more about the efficiency of engines and the interplay between their moving parts. Despite the progress that had been made, Gresley felt more could be done and pushed for a dedicated test facility, with rolling road and all the latest test equipment, to be built. Sadly, his arguments fell on deaf ears for many years, which clearly annoyed Gresley especially as just such a facility had been built in France. During a Presidential address to the ILocoE in 1934, he made his frustration at lack of progress in Britain only too clear, and sought

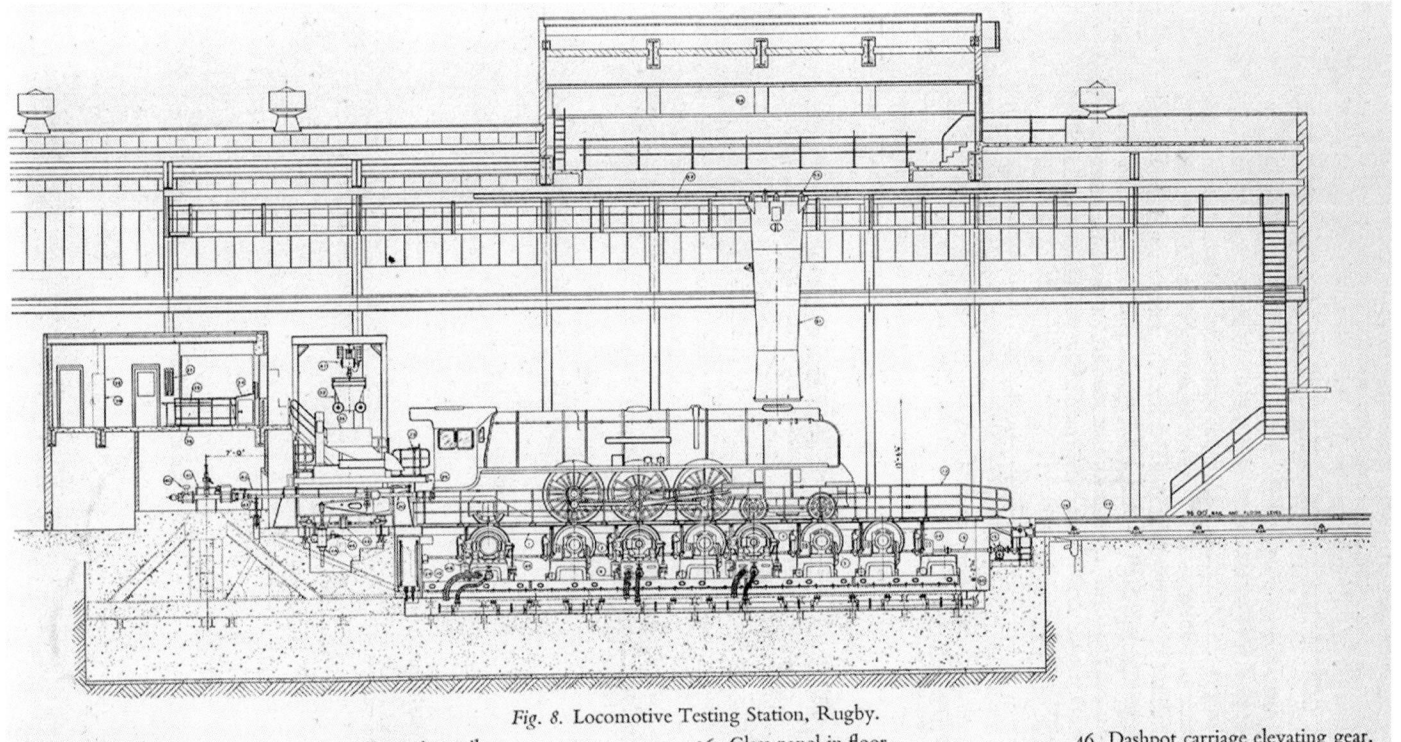

Fig. 8. Locomotive Testing Station, Rugby.

1. Lifting Platform.
2. Main drum.
3. Drumshaft bearing.
7. Dynamometer.
9. ,, roller.
10. ,, sole plate.
13. Removable gangway.
14. Crossbeam ,, support.
15. Reduction gear lifting jacks.
16. Operating shaft ,, ,,
17. Extension rails.
18. Rollers to withdraw platform.
20. Main drawbar.
23. Boiler feed control.
24. Injector overflow measuring tank.
25. Firing platform.
26. Coal hopper on weighing machine.
32. Portable coal truck.
34. Dynamometer control desk.
35. Amsler recording table.
36. Glass panel in floor.
37. Boiler feed tank telegages.
38. Gas analysis panel.
39. Pressure and temperature panel.
40. Amsler dynamometer.
41. Anchorage beam.
42. Mediating gear.
43. Rear drawbar support.
44. Damping dashpot.
45. Dashpot motor-operated needle valve.
46. Dashpot carriage elevating gear.
47. Electric travelling hoist.
48. Torque measuring gear.
49. Dynamometer dash pot.
50. ,, water supply main.
51. Chimney.
52. ,, sliding plate
53. ,, rollers
54. Smoke corridor.

Nigel Gresley's dream of a state of the art locomotive test centre was finally realised in 1948 when the facility for which he had canvassed since the 1920s was finally opened. It was hoped that the centre would add greatly to the business of design, but came too late to be of much use to supporters of steam. Sadly, with a few notable exceptions, including English Electric's GT3, the centre was not adapted to test diesel or electric designs where the future of the railways lay. It was here that it might have had a significant, longer term impact. (Author)

to shame the powers that be into affirmative action:

> About six years ago the French engineers, who were fully alive to the benefits which could be derived by the provision of a testing station, were able to persuade the French Government that it was necessary to have such a station [built]. Just over 12 months ago the French station at Vitry, near Paris, was opened.
>
> This experimental station is the most perfectly equipped in the world for carrying out analytical and scientific research….There are also four new dynamometer cars fitted with the most modern recording appliances, attached to the station. These are available for use in connection with trials in service on any French railways, and can be used for checking the results of innovations which have been introduced as a result of research in the locomotive experimental station…The establishment of the great new experimental station at Vitry is evidence of the confidence and conviction of the French engineers that progress can only be secured by full and complete research.
>
> What have we here in England? A small locomotive testing plant of 500 hp capacity, installed at the Swindon works of the GWR 30 years ago… The Swindon plant is, however, much too small for modern locomotives. (And) there are four dynamometer cars in existence on British railways, all of which I regard as almost obsolete when compared with modern cars.

To him and many others it was difficult to understand that as science advanced and locomotives became more complex, a testing regime that seemed more suited to the nineteenth century, should be their sole means of refining and perfecting new designs. Gresley was only too aware of the many variables in this process and the pressing need to understand them all fully, so that all possible outcomes could be considered and tested. And here, as I described in my book *Gresley and his Locomotives* (Pen and Sword Transport 2019), he:

> Looked to other emerging fields such as aviation and aerodynamics, which, during the 1920s, were reaching far into the future and pushing back boundaries. And they were aided in this by scientists working at RAE Farnborough, the National Physical Laboratory and major aircraft manufacturers such as Hawker and Supermarine. In these places advanced testing methods had become central to all they did and Gresley wished to bring this approach to the railways, where considerable amounts of money were being invested by the companies involved in new equipment programmes.

Despite clearly stating the operational need and strongly emphasizing the benefits, even the redoubtable and respected Gresley found the going difficult. Despite this, he persevered and the pressure he had been exerting for so long finally bore fruit. The design of a new test centre was eventually approved in 1938 with construction beginning shortly afterwards. But frustration was piled upon frustration when the coming of war a year later brought all work to a temporary halt and the centre only opened in October 1948, seven years after Gresley's death. However, its completion was celebrated widely by railway engineers, even though steam's demise was then being widely predicted so, perhaps, reducing the impact it might once have had.

Nevertheless, in 1948 it was seen as shining much needed light on the process of locomotive design and was greatly valued for that reason. For many years it was fully employed testing steam locomotives, though never made the switch to diesel or electric locomotion, where a longer-term role existed. There were notable exceptions though – English Electrics Gas turbine engine GT3, and two gas turbine engines produced by Brown Boveri and Metropolitan-Vickers for BR's Western Region. Then in 1965, the Brush-built Bo Bo diesel-electric engine, N0. 10800 named *Hawk* underwent tests at Rugby. But this wasn't enough to justify keeping such an expensive facility open and during the 1960s its role faded to nothing and the buildings were finally demolished in 1984.

As Britain recovered from the Second World War, locomotive designers and builders had many problems to overcome. First and foremost, there was a network requiring huge investment to restore its worn out infrastructure and a lack of money to do so. Then there was a nationalisation programme that hadn't given the expected boost to the development of new technology. And, finally, a seemingly overpowering desire of those in charge of BR to remain with steam for as long as possible, ignoring other options in the process.

At the same time, science, driven by the needs of war to develop new and better technologies, had encouraged designers to adapt emerging concepts to peacetime needs. So, there was an explosion of ideas coupled to better ways of testing, improving and measuring success. For many engineers and scientists, it must have seemed to be the beginning of a golden age to be explored and developed as quickly as possible. This was certainly a message promoted by the Festival of Britain during 1951 which served to engender 'one united act of national reassessment, and one corporate reaffirmation of faith in the nation's future'. Stirring words which seemed to capture the spirit of the age, but only time would tell if the country had the economic strength to turn these high blown words into reality.

Amongst the men and women recruited to fulfil this daunting commission, some found their way into railway engineering. The eager amongst them were driven by a need to apply new technologies to a steam dominated world and help bring Britain's network into the twentieth century along the way. One amongst them was John Hughes, who was born in Rotherfield Greys, near Henley on Thames in South Oxfordshire, on 6 February 1914. He was the second and youngest child of John, a Land Agent for the Earl of Derby, and Alice his wife.

The Festival of Britain in 1951 attempted to showcase the way new technology was being harnessed to take the country forward. British Railways occupied a less than prominent position in the exhibition halls, perhaps reflecting the gradual erosion of its status as the country's dominant transport system by road vehicles, and displayed some of its latest locomotives (left). The impression given was that new diesel and electric designs were taking precedence over steam, but this was very far from the truth. BR had, in fact, given priority to a massive steam locomotive building programme, with other forms of locomotion taking a back seat for the moment. (Author)

There is no record of anyone with a scientific bent in his family who might have influenced him in the choice of a career, but in a world where engineering was taking giant steps forward, particularly in the automotive and aviation industries, it isn't surprising that he was attracted to this world from an early age.

He was educated at Kelly College near Tavistock in Devon between 1925 and 1930, then became a tutor at the school for another two years. This was not an unusual practice in public schools at the time where gifted students were often deemed suitable to become stand-in teachers before moving on to university or entering a period of professional

Although John Hughes was fascinated by locomotive design, he could so easily have remained with Rolls Royce, where he had cut his teeth as a junior designer, developing cars. But for the war he may have done so. Yet his fascination with automobiles remained, so his wife Felicity later recalled, and he followed their development with much interest and enjoyed owning various types as this picture reveals. (JH)

training of some sort. In Hughes' case, this proved to be the Chelsea College of Aeronautical and Automobile Engineering in London where he began studying for a National Certificate of Mechanical Engineering.

During 1924 this independent college had come into existence, under the leadership of Sydney Roberts, a Cambridge graduate, with automotive engineering at its core. But with so much happening in aviation it was decided, during 1931, to expand its role to include aeronautical engineering, in the process establishing a very close link with the School of Flying at nearby Brooklands in Surrey. Hughes seems to have thrived at the college and graduated in 1935, with distinctions in Applied Engineering and Heat Engines. Shortly afterwards he was recruited by M.A. McEvoy's of Derby as a Test Engineer where he remained for a year. Whilst he undoubtedly gained much valuable experience in this post, it was not a role allowing him to apply his emerging design skills to any great extent. So in 1936 he applied to become a Junior Design Engineer with Rolls Royce's Car Division, spending the next three years helping develop the Phantom III and the 4½ litre Wraith and Bentley motor cars. Then war came and his career took a new path, as he later related:

On the outbreak of war I was transferred into the Design Office of the Rolls Royce Aviation Division concerned with carburation and engine control systems. During this time, I was involved in the development of a pilot's engine control system for fighter aircraft. I was then one of the inventors and designers of a fuel-injection pump which was used on the Griffon engine and the Merlin engine built in the USA for use with Mustang fighters. At the same time my duties were extended to include test bed and flight development programmes. As the war came to an end I led in designing the boost control used on high altitude versions of the Merlin engine. Much of this work gave me scope to consider and then submit a number of patents, which were accepted, with a number relating to gas turbines added later."

The war ending led to a slackening of the pace of development of aircraft, so Hughes returned to the automobile side of the Roll Royce business where he 'headed a section adapting an aero engine fuel injection system for use in Rolls Royce cars' for two years. But having been so closely involved in advanced design tasks, he looked for a post where his skills might be more fully employed and developed. So, in 1947 he joined English Electric becoming senior design engineer in its gradually expanding Gas Turbine Division. Here, part of the attraction was the company's increasing involvement in locomotive design. From early in his life, he had been attracted by railway engineering and like many youngsters at the time was a keen model maker. Now, as his career developed, this interest was re-awakened and with English Electric's increasing involvement in railway related projects, found a natural outlet in this field of research.

An Age of Opportunity and Change • 27

THE COLLEGE OF AERONAUTICAL AND AUTOMOBILE ENGINEERING
(of Chelsea)

Complete practical and technical training for entry to Civil and Commercial Aviation or the Automobile industry.
Entry from School-leaving age.

Syllabus from Bursar
Sydney Street, Chelsea, S.W.3.
Telephone FLAxman 0021.

Above and right: **John Hughes** would later write that 'Chelsea College of Aeronautical and Automobile Engineering not only equipped me with the skilled necessary to become a design engineer, but opened my mind to new areas of research that might be exploited in the future; turbines amongst them.' (Author)

NEW RESEARCH LABORATORIES OF THE INSTITUTION OF AUTOMOBILE ENGINEERS

CHASSIS TESTING DYNAMOMETER
ENGINE TESTING LABORATORY

GENERAL LABORATORY
MATERIALS TESTING DEPARTMENT

During the second half of the 1930s, when working in Derby, he had been close to the LMS's sprawling workshops there and followed very closely the advances in locomotive design being made by William Stanier and Tom Colman, his talented Chief Draughtsman. Later in life, he recalled the vivid impression made on him by development of the LMS's sole turbine Pacific locomotive, No. 6202, and remembered seeing it running along the West Coast Mainline on a number of occasions. He considered it:

> A valuable experiment that pointed the way to the future and one which the company, and then British Railways, should have pursued more vigorously. It had much potential to be exploited especially if and when it had been converted to burn oil not coal, which would have suited the turbine technology much better. I remember being very impressed when seeing it thunder by and the most distinctive sound it made as it did so. I was saddened when LMS, to all intents and purposes, abandoned the project before the war. It was clearly a missed opportunity.

When joining the Gas Turbine Division, Hughes was primarily involved in the development of a prototype 2700 HP lightweight gas turbine engine, designated EM-27. As is the case in many complex engineering projects, success was not immediately apparent. For several years English Electric struggled to make EM27 work effectively, though whilst doing so company planners thought deeply about all the possible applications it might have. Hughes began to take the lead in this work as he later recalled:

> In about 1953 I took charge of a project to assess how this engine might be applied to locomotive design and when

this was completed, and when my ideas had been accepted, prepared a case justifying the construction of prototype gas turbine locomotive with separate power turbine and mechanical transmission. This proposal took some time to be adopted, and there were many doubters, but eventually approval was forthcoming and funds were allocated, albeit limited. By this stage English Electric were heavily engaged in diesel development work and some saw turbines as a distraction from this potentially more lucrative area of work. This was especially so when BR's Modernisation Plan appeared 1955 with its emphasis on dieselisation and electrification. Against this background Turbines were seen as the poor cousin, though the Western Region's apparently successful experiments with two gas turbine engines in the late 1940s, early 1950s did add much weight to the case I made.

As a result of his energetic, well-argued advocacy of turbines, Hughes was appointed Chief Designer of 'Gas Turbine Locomotive Projects' in 1955 and controlled two design teams, one on the engines, transmission and auxiliaries based at Whetstone, and the other on frames, wheels, superstructures and more at the Vulcan Foundry, Newton-le-Willows'.

It had taken him longer than expected to reach this point, with many frustrating hurdles overcome along the way, but at last a project Hughes had long contemplated had the green light and could proceed. Recruitment of suitable specialist staff began and by the end of 1955 the serious business of creating a completely new locomotive began in earnest. Nevertheless, it was a project hindered from the start by a poor level of funding and a primary customer uncertain of its value and seeming to prefer a diesel and electric future.

Hughes was probably well aware of this and realised the problems he faced in bringing his work to fruition, especially having seen how other turbine projects had fared over the years in Britain and overseas. Most had struggled for success then quickly faded into obscurity when the promised benefits were not forthcoming, or companies stated a preference for other forms of motive power. Amongst his papers, Hughes kept details of all these projects and their outcomes., but he was not short of determination and undoubtedly believed from the start that he could succeed where others had failed or simply fallen short.

So, by 1955, when Hughes seriously began pursuing his dream of building a gas turbine engine that might conceivably change the course of railway history in Britain, there was much historic data to consider and many examples of success and failure to help chart his course.

In the early 1950s English Electric's managers, sensing that that gas turbines might find wider application in railway locomotives, set in motion a development programme that produced the EM-27. In John Hughes's hands this turbine would form a basic part of the company's GT3 project. When the EM-27 appeared it received wide coverage in trade journals and many brochures, such as this one, were circulated amongst portential customers. (Author)

Chapter 2
A SLOW BURNING REVOLUTION

A Turbine (taken from the Latin word '*Turbo*', which means to spin) – a machine for producing continuous power in which a wheel or rotor, typically fitted with vanes, is made to revolve by a fast-moving flow of water, steam, gas, air, or other fluid (*Oxford Dictionary*).

As is generally the case, a simple definition captures the spirit of a scientific idea but not its complexities. In principle a turbine, so simply defined in the *Oxford Dictionary*, contains a wheel or rotor which extracts kinetic energy from a fluid flow then converts it into mechanical energy. The simplest form of turbine is a windmill, which captures moving air in its sails and rotates them to drive a mill stone. In this most basic of examples, the primary parts of a turbine are a set of blades that catch a moving fluid, then a shaft or axle that rotates as the blades turn and finally a machine that is driven by the axle to produce an output.

Windmills, as we know them today, are believed to have first appeared during the eighth and ninth centuries in the Middle East. However, some historians suggested that the concept may have been explored in the 2nd millennium BC by the Babylonian ruler Hammurabi. He supposedly unveiled his plans to convert the power of the wind using an automated network of irrigation windmills to provide a regular flow of water to his land. True or not, it is not unrealistic to believe that using wind power to drive a machine occurred to early scientists who witnessed sailing ships in operation. But it would be many centuries before design engineers had the wherewithal to begin fully exploiting the turbine's potential. In my book *Turbomotive. Stanier's Advanced Pacific* (Pen and Sword Transport, 2017), I described this process of incremental advances in science and the factors that drove the speed of progress:

Turbine powered engines embraced a technology that had intrigued scientists for centuries, but had only been fully understood and applied in a practical way as the nineteenth century came to an end. As is frequently the case, such a development is dependent on the dynamics of different ideas, often spaced over long periods of time, coming together to produce a single advance of great significance – scientific dominoes falling to create a pattern often unseen by each individual participant.

The development of turbines, from these early examples, was a painfully slow business, simply

An early turbine designed by Giovanni Branca in the 17th Century. He was an Italian engineer and architect, believed by some historians to have designed but probably not built a basic steam turbine, except in model form. (JH)

Charles F. Brush built the first power generating wind turbine in Cleveland, Ohio during 1887. It was 4 tons in weight, rose to 60 feet in height, had a 144 blade propeller and a long tail. It generated just 12 kilowatts of electricity. (Author)

because potential applications could only be realised when other sources of power to drive them and stronger, more reliable materials with which to construct them became available. Industrialisation across many European countries in the nineteenth century provided the spur for more rapid progress. And here the steam engine was central to all that happened, particularly in the mechanisation of industry, movement of materials and people and the production of consumables to feed ever growing markets. As the century passed, and the societies touched by industrialisation began to undergo profound change, science kept studying the means of production and refined the drivers for this process and produced better, more efficient machines.

As the twentieth century approached, engineers sought to exploit alternative ways of powering this revolution. Turbines began to enter their vocabulary as one possible way of achieving this, encouraged by a growing understanding of thermodynamics, alternative power sources and ways of focussing energy to increase output. Wind, then steam, had been the primary driving force up to then, one came free and the other could only achieve a maximum isentropic efficiency up to 90 per cent. Gas burners, petrol driven internal combustion engines and the beginnings of electricity generation provided more efficient solutions. As understanding of these processes improved the potential of turbines began to be explored more fully, with steam initially leading the way.

As experiments increased understanding it was soon appreciated that steam driven turbines worked in two different ways – by impulse and the other by reaction. In the first of these, steam is forced at high speed through a narrow, fixed nozzle at bucket/cup-shaped turbine blades, these 'catch' the fluid and direct it off at an angle, making sure of constant energy impulses. This method increases the efficiency and strength of energy transfer.

In a reaction turbine, the blades sit in a much larger volume of fluid and turn smoothly and evenly as the fluid flows past them. The blades simply spin and any change of direction is small in comparison to that of blades in an impulse turbine. The rotor blades themselves are arranged and shaped to form convergent nozzles. To make sure that a steam turbine achieves its maximum efficiency, it was discovered that a mix of both impulse and reaction designs would help achieve this goal – the higher-pressure elements typically being reaction types, with lower-pressure stages using an impulse method.

In terms of development, it was not until 1884 that Anglo-Irish inventor Charles Parsons created a reaction steam turbine of real potential, which found wide use in the production of electricity in power stations and the development of high-performance ships' engines. At the same time, Swedish inventor Gustaf de Laval was building an effective impulse turbine and led the field in this alternative technology, in the process developing a nozzle that increased the steam jet to supersonic speeds (now named after him and still used in rockets today). Both his and Parsons's work was greatly influenced by the research and development work undertaken by the Slovakian scientist Aurel Stodola, who published two seminal works on steam and gas turbines, in 1903 and 1922. The collective results of these scientists' work were profound in the extreme and revolutionised the way machines were powered and their effectiveness.

By the beginning of the twentieth century, steam turbines had been chosen for use in power stations as an effective means of driving the generators that produced electricity. These facilities were initially driven by steam-powered reciprocating engines, but their size and slow speed made them increasingly

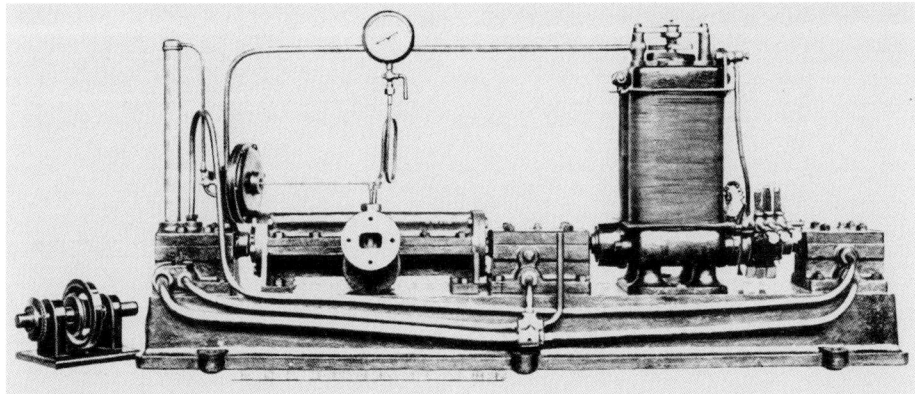

Three of the pioneers in developing steam turbines, in the process lighting the path for others, including John Hughes, to follow. (Above left) Charles Parsons demonstrating the workings of one of his turbines to the Prince of Wales in 1923. (Below left) His first turbine as it appeared in 1884. (Above right) Gustaf de Laval who led in producing an effective impulse turbine and, in the process, developed a new nozzle that increased a jet of steam to supersonic speeds. (Below right) Aurel Stodola whose research into turbines led to the publication of two seminal works that profoundly influenced their development. (Author)

uneconomic and steam turbines were seen as a very effective alternative. The world's first turbine powered facility – the Forth Banks Power Station – was commissioned in 1890 by the Newcastle and District Lighting Company, formed by Charles Parsons a year earlier.

Steam turbines also had a profound effect on marine propulsion and replaced larger, less efficient reciprocating engines. However, their advent was largely driven by the development of robust reduction gears, which could balance the revs per minute of turbines measured in thousands to propeller/shaft speeds of less than 300rpm, though some ships did manage with direct drive from the turbines to the propeller shafts (by directing the steam flow only).

The biggest and most dramatic impact of this new technology on ship design came with the launch of a turbine-driven 'yacht' laid down in 1894.

This boat was the product of another Charles Parsons enterprise, the Parsons Steam Turbine Company of Newcastle. Designed by Parsons, largely working in isolation, the boat

Power stations were the first major public amenities to benefit from the development of turbines. Here space was not a problem and engineers faced few restrictions when designing fixed, immobile machinery to generate electricity. Here a Metrovick built turbine, installed in Battersea Power Station, demonstrates this point only too clearly. Miniaturisation of such machinery, without sacrificing power output, for use in moving vehicles, presented designers with a considerable challenge. (RH)

made a dramatic appearance at the 1897 Diamond Anniversary Fleet Review, at Spithead, with the intention of achieving maximum publicity. The crew manoeuvred the boat at speeds of up to 34.5 knots in the limited space between the assembled ships, in what *The Times* later called 'a brilliant but unauthorised' display. The boat had been named *Turbinia*, in recognition of its source of power but also with an eye to headlines. By 1906, the Royal Navy was well advanced in using steam turbines, with other navies following suit, and this concept held sway in ship design through the two world wars until replaced by gas turbines, turbo electric drive and nuclear power.

As the developments of turbines in power generation and ship design evolved, their use began to attract the attention of some locomotive engineers. One of

Turbinia **undergoing** sea trials at high speed just before the end of the nineteenth century. Here, Charles Parsons managed to build a turbine small enough to fit into the limited space available in a hull without sacrificing power or creating resistance to the 'yacht's' high speed movement through water. The lessons he learnt here would soon be applied by locomotive engineers. (Author)

the main problems they faced was in building turbines small and light enough to fit into the limited space available. In ships and power stations this was much less of a problem, although the construction of *Turbinia* did require some thought being given to the need for miniaturisation to allow a working unit to fit into the hull of such a small 'yacht'. For railway locomotives, the benefits, on paper at least, seemed favourable, but in practice potential benefits and savings to be achieved over reciprocating engines might prove insufficient to justify mass production or continuation of experiments. Only time would tell if this was the case, and, of course, alternative sources power, such as gas or oil, might come along and, conceivably, offer greater benefits and efficiencies.

In the early years of the twentieth century there were a number of engineers prepared to explore the science of steam turbine powered engines, with Giuseppe Belluzzo in the vanguard of these designers. In so doing he became an expert in this area of science, writing some fifty or so publications on the subject along the way. Of these the most widely read was his 1905 '*Le turbine a vapore*'. In due course it became a primary teaching tool for engineering students around the world.

Such was the high esteem in which he was held that he was commissioned to lead a team charged with developing the first steam turbines for industrial use. These were later modified for use in ships. In 1906, he began applying all he had learnt to steam locomotives and designed the first turbine engine which was later built by the *Società Anonima Officine Meccaniche*.

This engine, which appeared in 1908, had two axles that were not coupled together and each axle was driven through a single mechanical reduction gear by two turbines. Even with uncoupled wheels, the behaviour and grip of the locomotive were thought excellent. Belluzzo continued his research but the coming of war in 1914 brought this work to a temporary halt. With austerity and a long period of recovery following this conflict, there was little appetite or funds to resurrect his locomotive designs in any meaningful way. However, this moratorium did not apply to power generation and his considerable talents found expression in this essential area of work. Nevertheless, when his increasing involvement in politics allowed, as he rose to an eminent position power in government under Benito Mussolini, he did become involved

Giuseppe Belluzzo was born in Verona in November 1876. Despite coming from an underprivileged background he attended college where he was awarded a degree in mechanical engineering. He then specialised in the design of turbines, in the process becoming a widely recognised authority on the subject. (Below) He soon turned his attention to the design of turbine powered locomotives with the engine shown here being his first working model. Belluzzo continued designing turbines for industrial purposes and, combining this task with that of politician, serving Benito Mussolini for many years in the process. (DN/Author)

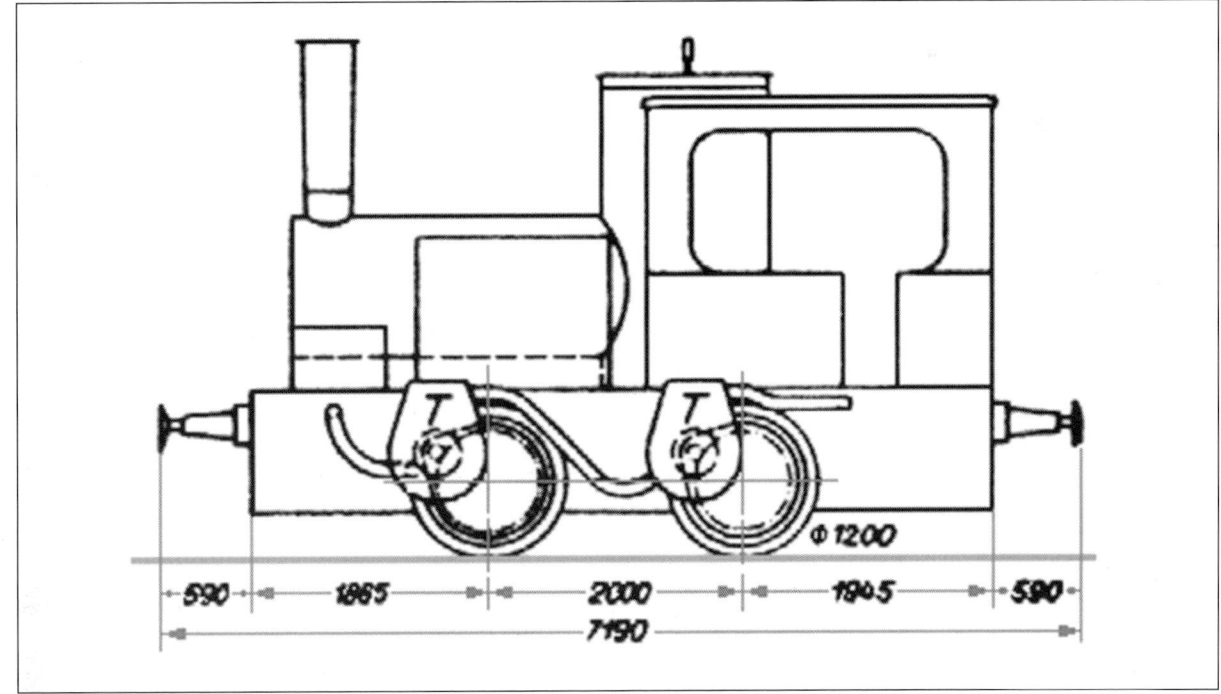

Right and below: **The Reid-Ramsey** turbo-electric condensing locomotive of 1910. (RH/Author)

in one more locomotive project. In 1931, Belluzzo acted as design consultant for a 2-8-2 turbine loco built by the Ernesto Breda company. It had high and low pressure turbines. It is thought to have been tested in the Breda works at Milan, but apparently the Italian State Railway would not allow it to be run on the main line, which effectively killed off the project.

Between these two Belluzzo experiments, other locomotive designers had taken note of his work with turbines and sought to adapt these developing ideas to their own regional railway systems. In England in 1910, Hugh Reid and D.M. Ramsey developed a turbo-electric design locomotive, of 4-4-0 + 4-4-0 configuration, weighing approximately 130 tons, for the North British Locomotive Company. It received little attention and trials proved inconclusive. In 1922 a second turbo-electric locomotive appeared, designed by Ramsey for Armstrong Whitworth. It was tested for eighteen months between Bolton and Southport primarily, pulling empty rolling stock. Its performance was poor, with consumption of fuel and water way above expectations. Testing continued into 1923, but despite modifications it did not improve and was withdrawn and eventually written off.

The next experiment, a geared turbine steam locomotive, was designed by Hugh Reid and James Macleod, a turbine expert, again for the North British Railway. On paper, *Challenger*, as the loco became known, seemed to hold out the hope of success and the prototype was completed in time for the 1924 British Empire Exhibition at Wembley, where it sat alongside the recently completed LNER Pacific *Flying Scotsman*.

This engine was built on continuous girder frames carrying a boiler, coal bunkers, water tanks, cab, condenser and auxiliary equipment all supported on two 4-4-0 bogies, with many parts coming from the company's 1910 experiment. Mainline trials did not start until 1926 and continued until April 1927 when it suffered axle box problems and then a serious turbine failure. It never ran again and so trials by North British were brought to an end. One can only assume that the expected benefits did not materialise.

As these experiments seem to confirm, turbines offered some potential benefits over traditional reciprocating locomotives. Soon, for example, designers came to believe that a steam turbine powered engine would be more efficient and consume less coal and water. In addition, their mechanical simplicity would make maintenance easier and improve reliability. In terms of wear and tear, traditional reciprocating engines were found to exert far greater downwards force than turbine locomotives. The strength of their hammer blows rose considerably so increasing the rate at which wheels and track wore out. The smoother take-up of turbine drive could reduce this, plus ease the problem of wheel slip considerably and, as an added benefit, make the ride on the footplate much less challenging.

However, even very early in development engineers foresaw some problems. They felt that greatest efficiency might only be achieved when pulling heavy loads and with the addition of condensers, which generated a vacuum to absorb the exhaust, but would add too much weight and significantly retard performance. Finally, it was believed that a single turbine could not be designed to run forwards and backwards,

After a pause of eleven years, Hugh Reid's (left as he appeared in 1921) second turbine condensing locomotive appeared (below), this time designed and built in partnership with James MacLeod and the North British Locomotive Company. It was a rebuild of the 1910 engine and used its frames, bogies and boiler, but on this occasion employed a direct drive turbine instead. Like Reid's 1910 prototype, it doesn't seems to have been a success, bringing an end to his turbine experiments. (RH/Author)

so a second unit was needed to make sure that an engine could reverse. Many other engineers in other countries believed these shortcomings could be resolved and development work continued, principally in Sweden, headed by the Ljungstrom Company, where new steam turbine locomotives were gradually evolving.

One such case was highlighted during 1926 in a paper presented to the Institution of Locomotive Engineers by J.G. Handley, Head of the Mechanical Department of the Argentine State Railway. In this he described a turbine locomotive bought by them to pull very challenging loads:

> The locomotive is designed to haul trains with a maximum wagon load of 700 tons at a speed of 25kph on long gradients of 1%, and the same load at a speed of 65kph if the gradient is only 0.4%. The locomotive is designed for running 800 kilometres without any renewal of its water or fuel, heavy duty oil. 'The whole performance of the engine may be considered satisfactory, apart from a few minor difficulties which were encountered.

In reality, the loco offered few advantages over more traditional designs, but engineers in Britain took note of the latest advance and watched developments at Ljungstrom, none more so than at the locomotive firm Beyer, Peacock and Co Ltd.

In 1926 a collaborative effort between these two companies resulted in another experimental locomotive being built, at a cost of £37,000. This engine ran for the first time in July of that year, followed by trials on the main line from St Pancras to Manchester and Leeds to London. In May 1928, the LMS

D.M. Ramsey, who worked with Hugh Reid before the Great War but didn't participate in his later turbine experiments, independently revisited the concept with Armstrong Whitworth's support. The company, having specialised in armaments work, were seeking to branch out into railway engineering, and saw turbines as being one way of breaking into this business, The engine (shown here) which they designed, and launched in 1921, proved, at 150 plus tons, to be too heavy and, during trials, displayed a lacklustre performance. By 1923, despite modifications, there had been no significant improvement and the project was dropped. (JH)

Drawing Office published the results for two of the Leeds test runs, where:

> On the whole the trains ran well to time. The principal loss of time which can be debited to the engine was due to two stops, totalling 27 minutes, made on 28th March on the 1 in 100 bank out of Sheffield due to a shortage of steam. This was caused by a damper having fallen shut in the air duct connecting the preheater with the ashpan resulting in defective combustion throughout the run. Coal consumption was considerably higher on the return run than on the outward trip and there is no apparent reason for this. When in good order the engine showed itself able to keep booked time in all sections.

During these tests, the engine developed a number of small defects, and these affected its fuel consumption and on some occasions its timekeeping. In comparison with a 2-6-0 engine, over the same route, the turbine engine used 4 per cent more coal. In water consumption, the saving over the standard engine was considerable, being 83.9 per cent on average. Once again, the tests showed that turbine engines ran relatively smoothly, but offered no great advantage over more traditional locomotives. But on this occasion the trials were spread over a longer period and gave the railways a chance to test the design more fully before rejecting it.

In reality these engines suffered from two handicaps that were hard to overcome. Most significantly there was a need to incorporate a condenser in the design. These were substantial in size and weight and very expensive to build and maintain. There was also a perceived prejudice, amongst running shed and maintenance staff, against such unconventional designs, so the workforce were naturally reticent in offering support. On the plus side, this design did add considerably to the ever-increasing knowledge of turbines in locomotives and pointed the way for future developments, with Ljungstrom continuing to take the lead.

Meanwhile in Germany, engineers, observing developments elsewhere, also experimented with turbine driven locomotives. Between 1927 and 1929 Krupp-Zoelly, Maffei and Henschel all produced their own models, but achieved only limited success, although two lasted in service until 1940 and 1943 respectively, whilst a third experimental engine was discarded in 1937.

Beginning in 1924, Beyer-Peacock of Manchester entered into a collaborative arrangement with Ljungstrom, the Swedish Company. Their chief designer, Frederick Ljungstrom, had been designing condensing turbines for a number of years and such a partnership seemed most advantageous. In 1926, the result of this work, a strange, ungainly engine, appeared (below). Trials with it continued until 1928 but were not deemed a success; the engine offering few, if any, advantages over reciprocating locomotives. (RH/Author)

German locomotive designers were also attracted by turbine technology in the 1920s and '30s, none more so that the partnership of Henschel, Krupp-Zoelly and Maffei. One of the key results of their work was the condensing engine shown to the right, which appeared in 1927. To save time and money, it was decided to convert a Class 38 4-6-0 locomotive by fitting a three stage forward turbine and a single stage reversing one. When under test its performance fell well short of expectations and, as a result, no more Class 38s seem to have been converted in the same way. Later on, Krupp-Zoelly developed a 4-6-2 locomotive that employed turbines built by the Swiss company Escher Weiss, but this project didn't go beyond the prototype stage. (RH/Author)

While this was happening, Ljungstom's designers had begun to focus on developing a non-condensing turbine locomotive and during 1931 came up with a workable solution. With Nydquist and Holm of Trollhattan they produced a 2-8-0 turbine locomotive to run alongside a number of 2-8-0 reciprocating engines operated by the Grangesborg-Oxelosund Railway. Stripped of the need to incorporate a condenser, this new locomotive had a balanced, more traditional look. It also weighed substantially less, allowing more of the generated power to be absorbed in pulling rolling stock, than wasted in coping with its own excessive weight. The engine proved a great success and three were built, remaining in service into the 1950s. They were seen as goods engines primarily, their usual and heaviest work hauling iron-ore trains of 1,750 tons up long inclines of 1 in 100. In service they ran more than 70,000 miles before needing a general repair, compared to the 36,000 miles achieved by the reciprocating 2-8-0s, with the added benefit of achieving a 23.8 per cent fuel saving. With three available, there was an economy of scale in procuring spare parts for the turbines and their ancillary equipment, the rest being shared

In 1931 Giuseppe Belluzzo again returned to theme of turbines being fitted to steam locomotives when acting as design consultant to the Breda Works in Milan. In his efforts to break into this area of research and development, having observed advances being made by engineers across Europe, Ernesto Breda sought Belluzzo's advice on how to proceed. However, the result of this private venture project (above), was not deemed a success and it was quietly dropped. (JH)

with the conventional 2-8-0s. So, in service they proved very effective and economical to run and were only replaced when diesel locomotives were eventually introduced.

It was the development of these three engines in Sweden that influenced William Stanier when considering building a turbine locomotive for the LMS. This long held ambition was fostered, in part, by his old friend and fellow engineer, Henry Lewis Guy, who was a noted innovator and inventor. Born in Penarth in 1887 he became an apprentice with the

(Left) **Fredrik Ljungström**, a Swedish engineer, was considered to be one of the foremost inventors of his time. As a result he was able to generate hundreds of technical patents during his career. He was also an industrialist of great note who early in his life saw the possibilities inherent in steam generator and turbine technology. As a result, he sought to exploit their power in a host of designs, including power stations, boats and locomotives. After a long and successful career, he died in 1964. (Below) Perhaps the most successful turbine locomotive of the 1920s and '30s was the Ljungstrom/Nydquist and Holm built 2-8-0 design. Deciding that a non-condensing type of locomotive had many advantages over a condensing version, this partnership invested heavily in this alternative technology. The result was a class of three engines for use in pulling long, heavy iron-ore loads over the challenging line run by Grangesborg-Oxelosund Railway across Sweden to the Baltic Coast. In this role they proved highly successful and remained in service well into the 1950s, finally being replaced by diesel locomotives. All three engines have been preserved. (RH/Author)

Taff Vale Railway before studying mechanical, civil and electrical engineering at the University of Wales. By 1918 he had risen to become Chief Engineer to Metropolitan-Vickers where he put his considerable skills into the development of turbines to good use. As friends and fellow members of the Institution of Mechanical Engineers, he and Stanier were able to establish a strong professional bond. However, it wasn't until Stanier became the LMS's Chief Mechanical Engineer that they were able to pursue the idea of developing a turbine powered steam locomotive with any real chance of their proposals being accepted.

Guy and Richard Bailey, a fellow scientist at Metrovick also with a keen interest in turbines, had been close associates of Frederik Ljungström for many years and their great experience of turbines had proved invaluable to the Swedish company when designing their non-condensing engine. So when word of its success began reaching them, in late 1931, they were eager to see how it worked and judge for themselves if Ljungström's solution might be adapted for use in Britain. As they studied the results of trials in Sweden, they engaged Stanier, who moved to the LMS in January 1932, in this process undoubtedly eager to enlist his support in making the project a reality. But the process was not a quick or easy to manage, with progress largely dependent on Stanier being able to meet the company's broader development needs in terms of new locomotives and rolling stock. This didn't happen overnight, but soon sufficient progress had been made to cement his reputation and so allow him a certain amount of freedom to experiment. And so in late 1932, Stanier, accompanied by Guy, Bailey and Harold Chambers, his own Chief Draughtsman at Derby, visited Sweden to see for himself the new non-condensing turbine locomotive in operation.

Armed with all the background information they needed, Stanier and Guy submitted papers to their respective boards seeking approval to build a single experimental turbine locomotive. After brief discussions, their proposal was endorsed in February 1933, with the engine, a Pacific, being added to the Princess Royal programme which had been approved the previous year. With such a cutting-edge project to manage, the need for research and development to iron potential problems was unavoidable, with one key issue being the way the engine would be controlled. Guy and Bailey believed that 6202 would be best served by having a single turbine capable of both forward and reverse movement. This was an issue widely debated, because experience to date had shown that a single turbine was most effective when only able to work in one direction. With this in mind, it was suggested by some that 6202 should have two turbines, not one. Guy thought otherwise and intimated that a two-way function could be built into a single turbine. But when asked how long this would take to design, could not say with any certainty.

Having no wish to delay the project or move away from a concept that was already known to work and take a potentially expensive, unproven route did not appeal to Stanier, or the company chairman Josiah Stamp for that matter. Nevertheless, it was a debate that Guy and Bailey revisited over the years as problems continued to bedevil the reverse turbine.

Work on the engine was completed in June 1935 at Crewe and that month she began a prolonged period of testing. From the beginning of her service, she was based at Camden Shed and would ply her trade on the West Coast Mainline until rebuilt as a conventional engine in 1952. And it was in this form that the engine was badly damaged in the Harrow and Wealdstone rail disaster later that year, written off and then scrapped. There is no denying this engine's important place in railway history and then there are the inevitable 'what might have been' elements to reflect upon. Yet, undeniably, there is one other issue to address – the effect this project had on John Hughes in his formative years as a design engineer and the way Stanier and Guy's great experiment influenced the course of his own career.

Although steam continued to dominate locomotive design, in the years after the Second World War, its grip was slowly easing as new technologies, based on the internal combustion engine, slowly took hold. Looking back, it seems that some engineers held on to steam for far too long, while the economic value of cheap coal in near bankrupted countries proved too tempting for politicians to ignore. Nevertheless, a push towards a diesel and electric future, though lacking the impetus it deserved, did begin to take shape, though Britain tended to lag behind other Western democracies in the speed of change. In such a situation, turbines were unlikely to find immediate favour, especially when other tried and tested alternatives were becoming available. So the main choice was between dieselisation and electrification. If Stanier's turbine Pacific had been built in greater numbers, or a planned 2-8-0 had appeared, things might have been different. But these plans died, and turbine development took a back seat as other options were pursued.

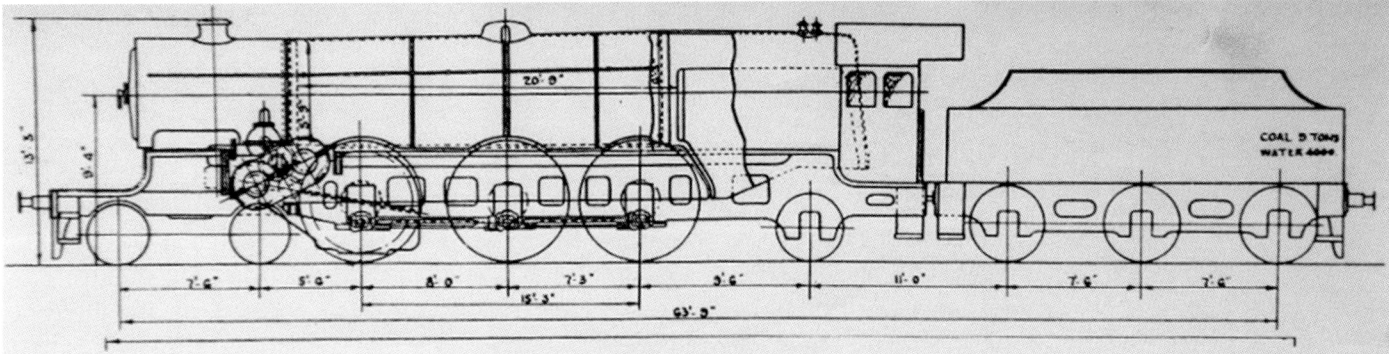

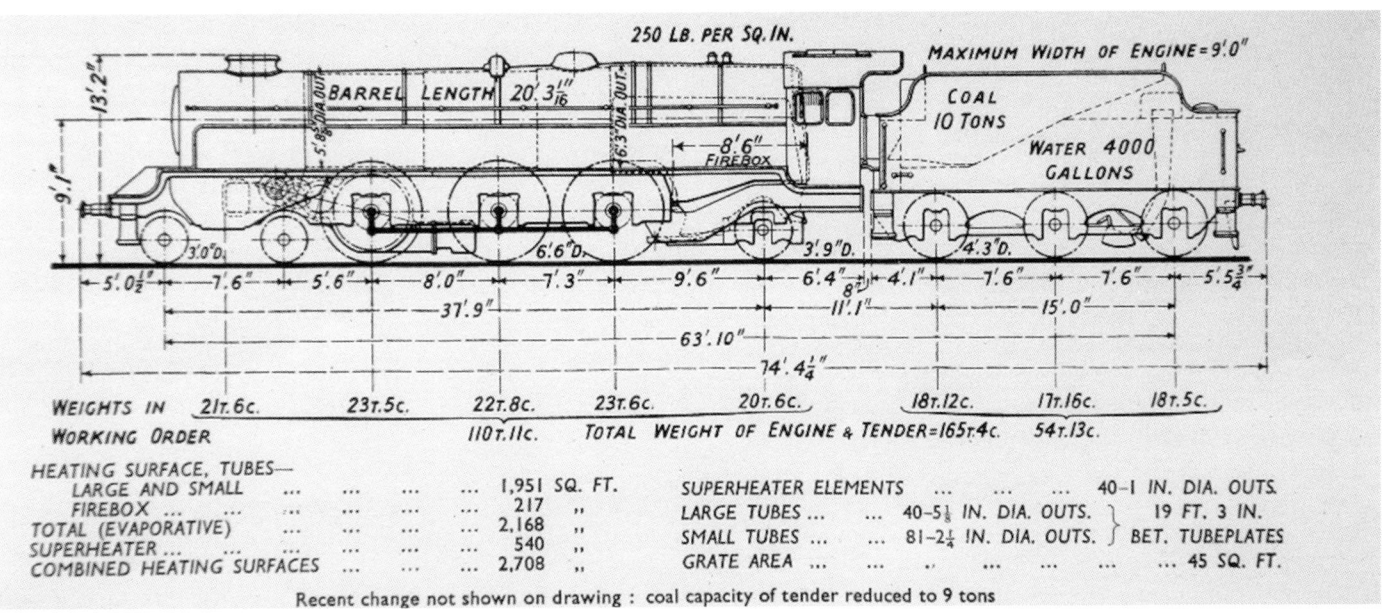

Although by 1932 various experimental turbine engines had been built in Britain, none had found favour. Nevertheless, William Stanier of the LMS (top left) and Henry Guy (top right) were convinced that, with the construction of a successful non-condensing locomotive in Sweden as an example, something similar was possible in the United Kingdom. As a result, approval to build a turbine locomotive based on the new Princess Royal class was given in early 1933. However, with many complex design issues to be resolved, the engine, numbered 6202, went through a number of evolutions before it was finally turned out by Crewe Works in June 1935. The top diagram shows the Derby Drawing Office's first attempt at designing their turbine locomotive in 1932/33 as Stanier sought the LMS's approval to proceed. The lower diagram shows 6202 when built. Over the next ten years she underwent a number of modifications to help enhance her performance, which, in any event, was considered good (JH/RH/Author)

Right and below: **June 1935 and 6202**, soon nicknamed *Turbomotive*, begins a prolonged period of testing. Although deemed a success, the LMS didn't build anymore steam turbine engines of any type and in 1952 the experiment drew to a close with the engine rebuilt as a conventional Pacific. (Author)

Nevertheless, there were still some prepared to experiment with gas turbine technology as various experiments during this period would show.

Before describing the evolution of gas turbine locomotives, leading to GT3 and beyond, it is interesting to consider how it compared to the steam concept explored by Stanier, Guy and Ljungstrom amongst others in the 1920s and '30s. In an early development paper, produced by John Hughes for senior English Electric managers, he sought to explain how the two systems differed:

> A steam turbine adopts a process that extracts from pressurized and uses it to do on a rotating output shaft. By comparison a gas turbine is,

essentially, a combustion engine within a power plant that is used to change the natural gas or fuels of liquid to mechanical energy. Here it must be realised that gas turbines operate on an open cycle in which air is taken from the atmosphere, then compressed in a centrifugal or axial-flow compressor. This is then fed into a combustion chamber where fuel is added and burned at a constant pressure with a portion of the air.

Reciprocating engines depend on the to and fro motion of a piston, which must then be converted to rotary motion by a crankshaft arrangement, whereas a gas turbine delivers rotary shaft power in a smoother, more efficient way. In essence this allows a considerable amount of power to be produced by such an engine for the same output. In addition, it will be smaller in size and lighter than a reciprocating engine and so achieve a better power to weight ratio. The advantages to be gained from this over steam turbine locomotives are readily apparent. There may also be clear advantages over the diesel and electric locomotives gradually entering service, but this can only be determined by a series of controlled, comparability trials."

When writing this paper, in the immediate aftermath of the war, Hughes was clearly aware that gas turbines had developed a long way from a concept first investigated by John Barber in the eighteenth century. His proposals had identified all of the important parts of a such an engine, including a chain-driven reciprocating gas compressor, a combustion chamber and a turbine. Barber's aim in pursuing these ideas was driven by a need to find a way of propelling a 'horseless carriage'. In time, he worked his ideas up into a Patent (No. 1,833, which was published in October 1791, a copy of which Hughes kept). The device he designed had two retorts (devices used for distilling substances) which created gas from coal, this was then burnt and passed through a nozzle to impact on an impulse-type turbine. Before combustion, the air and gas were compressed by a piston-type compressor driven from the turbine. In this example, provision was made for injecting water into the burning gases 'to prevent the inward pipes and mouth of the exploder from melting by the intensity of the issuing flame.'

For the time such a far-seeing invention had applications that even engineers in the early twentieth century could barely

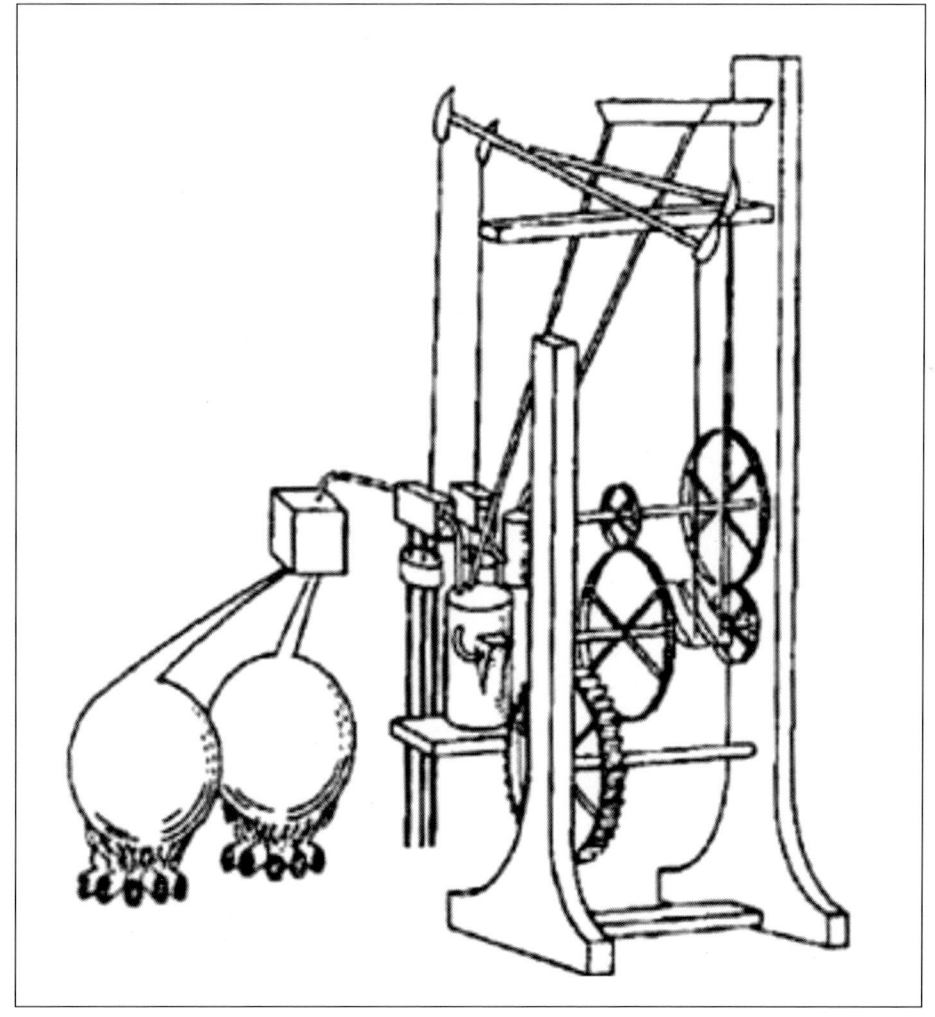

The sketch contained in John Barber's 1791 patent No. 1833 for a gas turbine engine. Here two retorts (left) are heated to create distillilation and produce gas. This then gets fed into the receiver and from there into a combustion chamber and receiver to power the turbine. (JH)

perceive, the jet engine being the most remarkable. In the 1790s, given the unrefined state of science at the time, it wasn't possible for Barber's invention, no matter how good, to generate the power needed to compress air and gas and produce a worthwhile output for the vehicle he planned. Nevertheless, as John Hughes readily appreciated, this invention provided the impetus for designers to slowly refine the concept. And gradually its potential was unlocked as the materials and wherewithal to create a machine strong enough to survive the pressure and stresses involved were finally developed.

For railway engineers eager to exploit gas turbines two versions emerged during the 1930s and '40s – gas turbine-electric and a gas turbine-mechanical. In the first of these a turbine is used to drive an electric generator or alternator, so producing a current sufficient to power a traction motor. In the second, a mechanical transmission is employed to deliver turbine power directly to the wheels. As each type was developed their benefits and limitations soon became apparent. For example, gas turbine-electric locomotives were found to be very noisy and tended to produce extremely hot and potentially damaging exhaust gasses. Then there was the question of power output and efficiency, which dropped with the frequency of rotation, to consider. Here piston engines, which have a comparatively flat power curve, could boast a clear advantage. To overcome this shortcoming gas turbine-electric locomotives were deemed better suited to long-distance high-speed runs.

When it came to assessing the gas-turbine mechanical option, John Hughes, during 1947, in a submission to his managers, provided a very clear view of the good and the not so good elements, though he may have played down the shortcomings. This is, perhaps, unsurprising when considering the choices he would soon have to make in designing and building GT3:

It can give high power with compactness and light weight, together with excellent reliability and a reduction in complication compare with other prime movers. The torque at standstill may be made several times that at the design speed so that it is practicable to use a simple mechanical coupling to the wheels with appreciable gain in tractive effort at speed, which is very appropriate for line service. The number of rubbing parts are few and the consumption of lubricating oil very small. It required little maintenance and gives high availability and offers the possibility of burning lower grade fuels.

He also suggested one other less quantifiable advantage, in a separate paper to English Electric's senior managers. In it he wrote that:

It seems that British Railway officials, who wish to retain steam locomotives may find it easier to accept a turbine-mechanical and not a diesel or electric alternative. This is because, although not powered by coal, it can be built in such a way as to contain many mechanical features of a steam locomotive – frames, wheel configuration, motion and so on. When selling a new type of engine to BR this may give us a distinct advantage over our competitors. There is also the cost to consider. By using existing frames, wheels and motion savings can be made in the overall outlay on production.

Perhaps this might, with the benefit of hindsight, be considered a naïve view, but it does at least reveal Hughes's desire to get his gas turbine locomotive built and what he thought he had to do to achieve this ambition. If so, he succeeded up to a point, but then found himself committed to building a new locomotive too wedded to the past. At a time when railway engineers were finally embracing the new world of diesel and electric locomotion this coy reminder of a soon to disappear technology stood in sharp contrast to what was then happening across the network. So, a possible advantage ultimately became a potential weakness.

However, there were other identifiable shortcomings as Hughes's 1947 paper makes clear:

The disadvantages are that efficiency of the simple cycle gas turbine falls off substantially with a reduction from full power, and until recently the cost per horsepower was higher than for other prime movers.

There is also the question of the high costs and risks involved in developing a prototype in a barely explored field of railway engineering. In research and development projects companies have to gamble much on possible outcomes, the effectiveness of the product and then being able to find a market. High, speculative costs are loaded in the early years of a project, which, and even if successful, may never earn enough to cover this investment. The higher and more advanced the technical specification the greater the risk. Only with very large orders and continuous mass production can financial success become possible. With the costs of the two world wars, and years of recession between times, virtually bankrupting our nation all industries tend to avoid risk.

This is unlikely to change except by Government investment – which for the foreseeable future seems unlikely – or investors being attracted from overseas, especially the United States."

By the time Hughes was penning these words, other locomotive designers and companies around the world had indeed begun to invest in, amongst other things, turbine technology. So by the late 1940s, early 1950s there was a growing bank of material he could consult when considering his own project.

The first in the field with a gas turbine-electric locomotive was the Swiss Federal Railway. In 1939 its managers, eager to experiment with this new technology as a means of meeting a specific operational

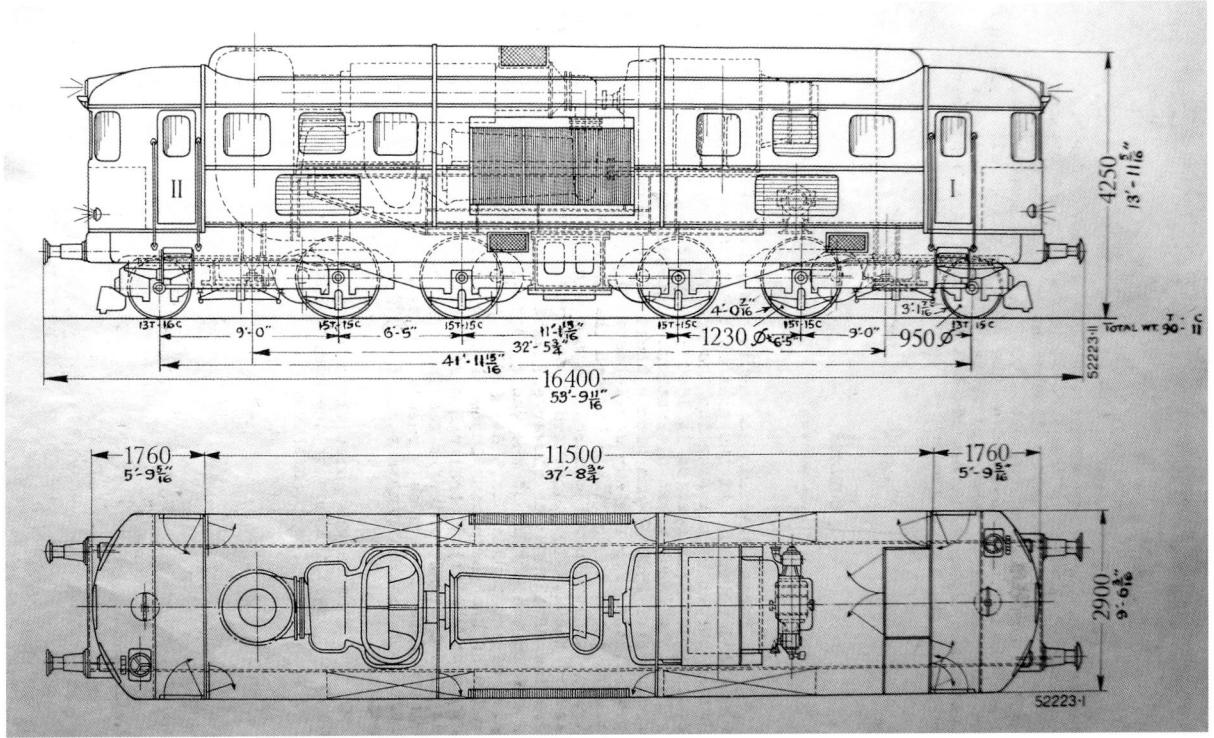

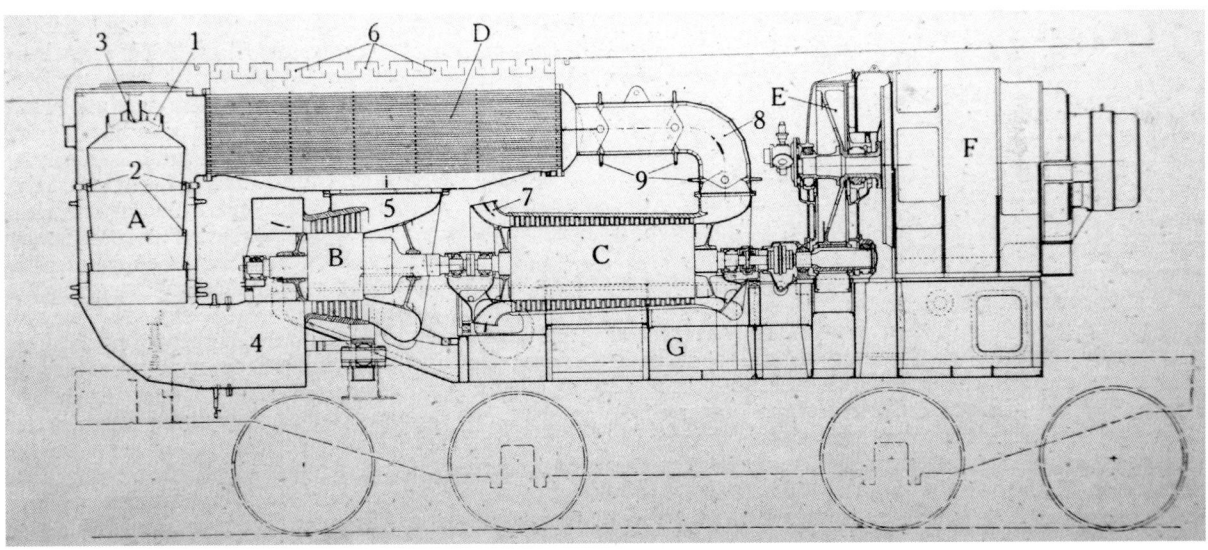

(Top) **The drawing** prepared by designers working for Brown-Boveri in 1940 showing the general layout and dimensions of the gas turbine-electric locomotive (Am 4/6) they were commissioned to build for the Swiss Federal Railway. (Above) A more detailed drawing showing the general arrangement of Am 4/6's gas turbine The parts are listed as: A. The combustion chamber. B. The gas turbine. C. The compressor. D. The gearing. E. The generator. F. The bed plates. 1. The swirl vanes. 2. Slits. 3. The nozzle. 4. The burner. 5. The exhaust passage. 6. Slits. 7. The inlet. 8. Outlet pipe. 9. The expansion joints. (JH)

need, ordered a single gas turbine engine – one capable of, producing 2,170 horse-power - from Brown-Boveri. It was completed in 1941, and then underwent testing before entering service.

The Am 4/6, as it was called, was required to meet the need for locomotives to pull fast, passenger trains on low density routes where electrification was deemed to be uneconomic. From May 1943 to July 1944 the locomotive was trialled on the Winterthur-Stein-Säckingen line where it covered 150 km daily pulling mixed loads of up to 300 tons. Within a year it had covered a total distance of around 50,000 km without experiencing any serious difficulties.

With war raging across Europe news of this development was slow to spread. Nevertheless, Hughes, by some means in 1944, obtained a Swiss magazine article describing the engine and its performance during the trials. Then in February 1946 the *Locomotive* magazine ran a detailed account of the engine's development, much of which Hughes highlighted in red ink, even though he was still a year or more away from joining English Electric's turbine division. The key points he noted were:

Some idea of the importance attached to this development may be gathered from the fact that notwithstanding the exceptional difficulty in obtaining fuel oil during the war, the federal Government released to the railway sufficient to put the locomotive into regular service … it is gratifying to be able to report that the results have so far proved entirely satisfactory.

In order to develop a useful output of 2,000hp the gas turbine must develop an output of 8,000hp, the compressor absorbing 6,000hp. To transmit the available power to the driving wheels the designers' had the choice of mechanical, hydraulic, pneumatic or electric transmission – or a combination of these methods. Direct transmission was excluded for a number of reasons. Firstly, it would result in the speed of the gas turbine and compressor unit varying directly with that of the locomotive, an inadmissible arrangement even if gearing were provided. Secondly, the reduction gear, with a ratio of the order of 10 to 1 would have to allow the set to be coupled and uncoupled while running - the lower power unit having to be run up light and engaged afterwards.

Gear-changing and clutch devices for such high powers have not yet been tried out in practice. The same applies to hydraulic and pneumatic transmission systems, which might overcome the first mentioned difficulty. It is true that numerous hydraulic transmissions have been successfully applied to locomotives of up to 400hp, but satisfactory operating experience with considerably higher powers does not appear to be yet available.

The acceleration (of Am 4/6) is both smooth and rapid, which is attributable to the

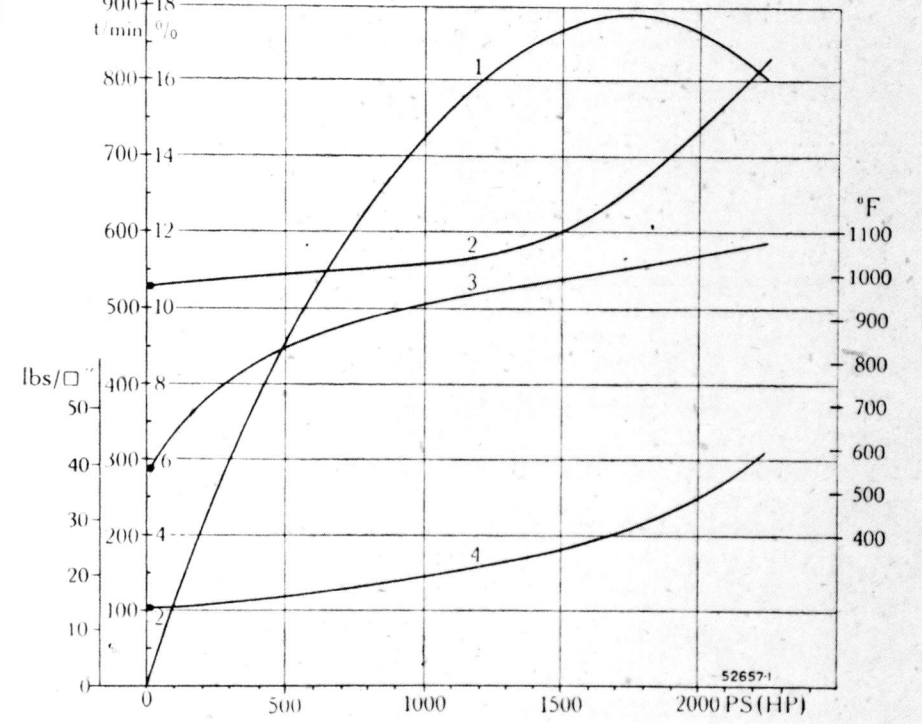

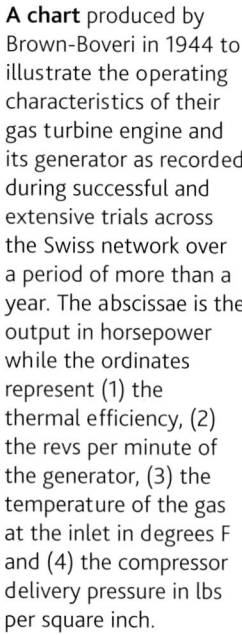

A chart produced by Brown-Boveri in 1944 to illustrate the operating characteristics of their gas turbine engine and its generator as recorded during successful and extensive trials across the Swiss network over a period of more than a year. The abscissae is the output in horsepower while the ordinates represent (1) the thermal efficiency, (2) the revs per minute of the generator, (3) the temperature of the gas at the inlet in degrees F and (4) the compressor delivery pressure in lbs per square inch.

In 1939 senior managers in the Swiss Federal Railway, having been impressed by some conceptual work undertaken by Brown-Boveri, an electrical engineering business established by Charles Brown and Walter Boveri in Zurich during 1891, ordered a single gas turbine electric locomotives for experimental purposes. It was completed in 1941 and is seen here when under test during 1942. It proved to be a sound design but Brown-Boveri were not commissioned to build any others for the company. (JH)

General Electric's attempt to build a steam turbine locomotive for Union Pacific produced a class of two in 1939 (one of which is shown below). These 2-C+C-2 engines were placed under test for six months but were deemed to be unsuited to the company's needs and were returned to GE the same year. The failure of this experiment and UP's desire to explore the potential of gas turbines led to the development, by GE, of the more successful UP 50s. (JH)

high inertia of the rapidly rotating masses of the power unit. When more power is required the control wheel is rotated a few notches further, the governing gear automatically making the several adjustments necessary.

The Allis-Chalmers Manufacturing Company of Milwaukie, a licensee of Brown-Boveri Ltd, had announced its intention of building a 5,000hp gas turbine with mechanical-hydraulic transmission and the operating results of this locomotive are awaited with interest.

It is unclear whether John Hughes, during the hiatus on travel caused by the war, was able to visit Switzerland to see for himself Am 4/6 in action. However, his archive contains a number of photographs of the locomotive in day to day action, to sit alongside a mass of technical information which undoubtedly came directly from Brown-Boveri in 1946/47. And in his papers, he noted the progress being made during tests and the engine's eventual fate.

For a decade or more Am 4/6 was used in experimental service across Switzerland, France and Germany. During this period, she covered some 410,000km. Then in September 1954 she was damaged when the turbine suffered a bout of severe overheating caused by faulty wiring of sensor cables. Due to the cost involved in repairing the extensive damage caused by this incident and the relatively poor fuel efficiency of the locomotive, it was decided to end the experiment and withdraw the engine from service. Subsequently, Swiss Federal Railway officials decided to rebuild her as a purely electrical engine, reclassifying her Ae 4/6 III 10851 in the process. In this state the locomotive remained in service until 1978.

By the early 1950s there were other gas turbine projects reaching fruition that caught Hughes's attention, judging by the papers he kept in his archive. Three of these proved of particular interest to him – two in Britain and the third in the United States.

During the late 1930s, General Electric began experimenting with steam-turbine-electric technology to meet a need for more powerful passenger locomotives to run over the Union Pacific's lines. Two were developed over three years at the company's works in Erie, Pennsylvania, at a cost in the region of $2million. Designed to resemble the Electro Motive Diesels, with which they might work in tandem, they entered service in 1939. The wheel arrangement of both locomotives was 2-C+C-2 and their engines produced 2,500hp and as such were able, in principle to pull heavy loads up to a maximum speed of 125mph.

After several trial trips over the New York Central network, they left the Erie works in April destined for Omaha to join the Union Pacific's fleet. After a proving period, they were assigned to duties that included pulling streamlined passenger trains across the west region of the country. However, they were only in service for 6 months or so before being returned to General Electric as being unsuited to their needs. And so a search began for an alternative.

General Electric had been building gas turbines for aircraft and proposed using something similar on a locomotive. Based on a limited knowledge of the technology and the power it could produce, Union Pacific's managers, in an effort to minimize the risks and costs involved, decided that such a locomotive, if it were to realize its potential, should focus on mainline freight duties and equal the power of the steam engines deployed to this task. Here, it was concluded, working demands and relatively high speeds required would maximize the turbines' efficiency.

With an expression of interest from Union Pacific in their pocket, General Electric proceeded to build a prototype, No. GE 101, which they completed in November 1948. After completing tests across the northeast during June the following year, it was taken into service, though not ownership, renumbered UP 50, painted in the company's yellow livery and soon began a series of ever more demanding trials to test its capabilities and suitability.

This class of engine was built to a car body design. In such an arrangement a bridge-truss framework is used so making the body a structural element of the locomotive. Such a design gained hugely in strength because of the rigidity of these frames and was noted as requiring far less structural weight to achieve rigidity than do locomotives with non-structural bodies. For that reason, car body construction was favoured as it was deemed to produce a better power-to-weight ratio than early diesel locomotives. In UP 50, to increase its structural strength even more, the body was extended down its full length and internal walkways

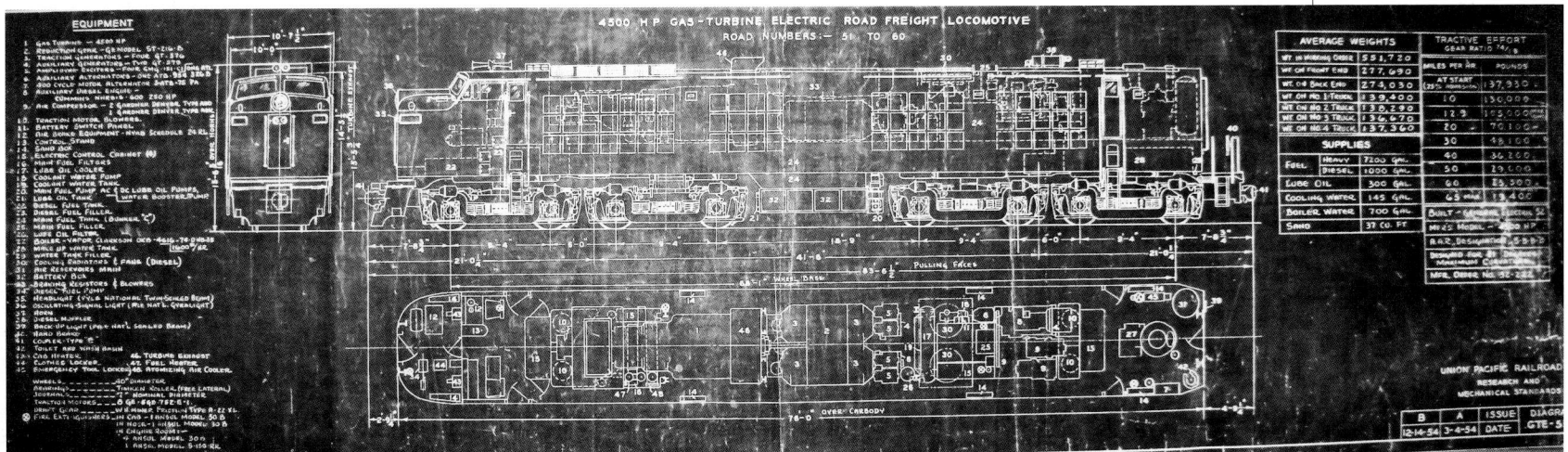

A blueprint of the General Electric substantial gas turbine locomotive built for the Union Pacific Railroad in 1948. This drawing was acquired by John Hughes when both UP and GE sent him a large package of performance reports and photographs relating to the project in 1949. (JH)

Fig. 1 Pilot gas-turbine-electric locomotive ready for field tests

Fig. 2 First design of commercial gas turbine-electric locomotive

Fig. 3 Second design of commercial gas turbine-electric locomotive

Fig. 4 Model showing general appearance of the 2-unit, 8500-hp locomotive currently under construction

John Hughes carefully collected material relating to the General Electric built UP 50 gas turbine engines. The first of these, a prototype, appeared in 1948 and extensive testing led to a three-stage construction programme lasting from 1952 to 1961. This series of photographs shows how the design was developed. They were deemed a success, but their fuel eventually proved more expensive than diesel equivalents and replacements were sought in the early 1960s and these began arriving in 1963. (JH)

inside the framework were included to make servicing easier. The prototype was also fitted with cabs at both ends to 'help speed up turn around times' as the company's publicity material put it.

UP 50 weighed 230,000kg and was over 24m long. A B+B-B+B wheel arrangement was favoured with four two-axle trucks with pairs connected by span bolsters. Its turbine was rated to produce 4,800hp of which 4,500hp was available for traction. This power output was more than double that of diesel-electric units at that time, which is why Union Pacific were prepared to gamble on such unrefined technology. In due course diesel designs would improve considerably so closing the performance gap.

After a period of successful testing, Union Pacific officials began ordering the engines in bulk primarily as a means of replacing its twenty-five Big Boy steam locomotives. The fifty-five engines that made up the new class were delivered in three batches between 1952 and 1961, with each generation suitably modified as operational experience grew. For example, a single cab was fitted which freed up space for more fuel to carried, higher output engines and in 1953 there was even an unsuccessful experiment with UP 57 which was converted to burn propane gas fed from a pressurized tank carried in a tender.

Generally speaking, these engines were thought to be successful and served the Union Pacific well for many years. Nevertheless, their fuel consumption was thought to be poor – the turbines consuming roughly twice as much fuel as a comparable diesel engine. This was initially not considered to be a significant problem, because their turbines burned Bunker C heavy fuel oil, which was less expensive than diesel. But this highly viscous fuel is notoriously difficult to handle. As diesel technology improved, and these

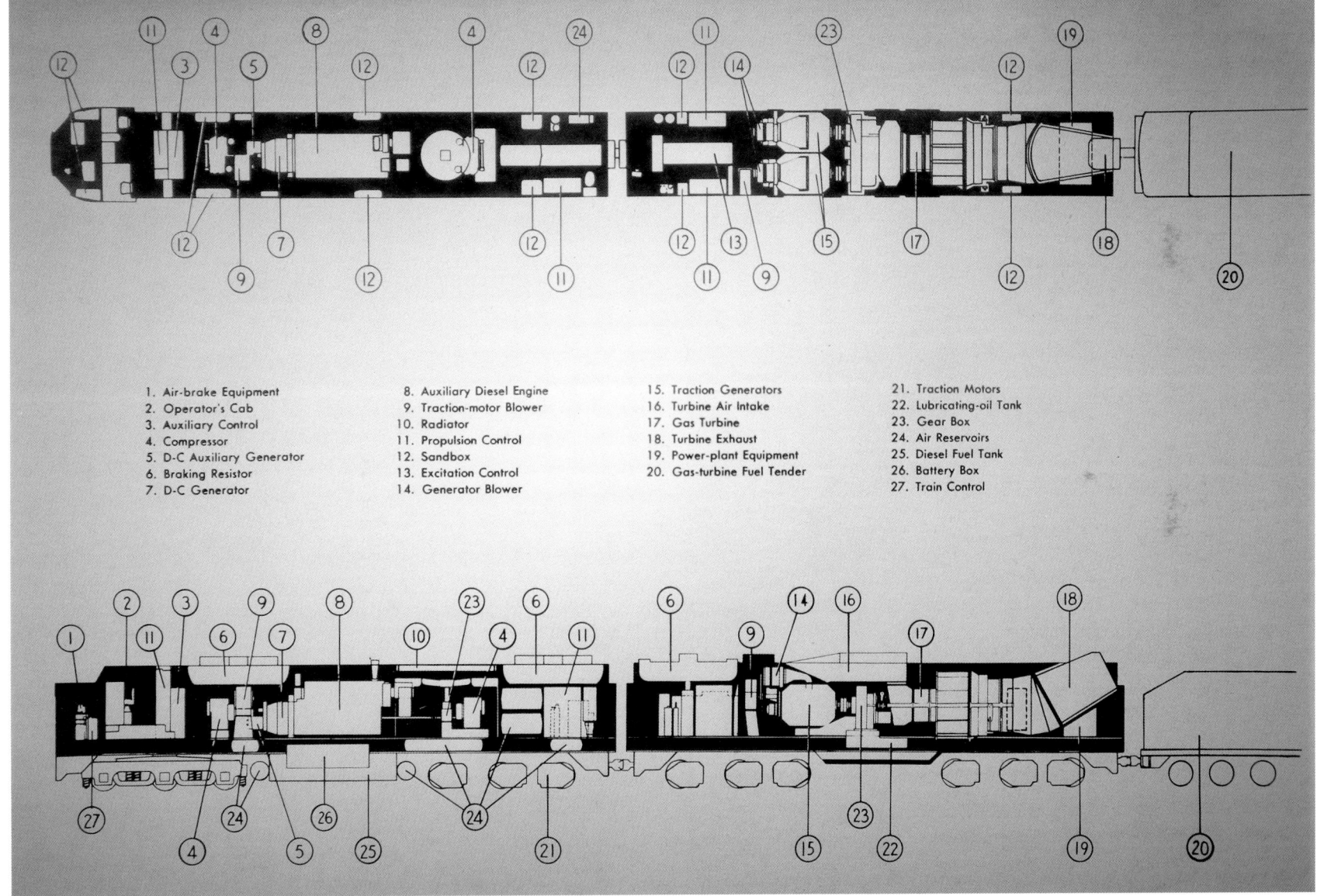

1. Air-brake Equipment
2. Operator's Cab
3. Auxiliary Control
4. Compressor
5. D-C Auxiliary Generator
6. Braking Resistor
7. D-C Generator
8. Auxiliary Diesel Engine
9. Traction-motor Blower
10. Radiator
11. Propulsion Control
12. Sandbox
13. Excitation Control
14. Generator Blower
15. Traction Generators
16. Turbine Air Intake
17. Gas Turbine
18. Turbine Exhaust
19. Power-plant Equipment
20. Gas-turbine Fuel Tender
21. Traction Motors
22. Lubricating-oil Tank
23. Gear Box
24. Air Reservoirs
25. Diesel Fuel Tank
26. Battery Box
27. Train Control

On the back of this diagram John Hughes has written 'UP's gas turbine engine in its final form with second car and fuel tender to increase its range over those long drags so common in the States – very impressive'. There appears to be no record of him visiting the USA to see them in operation, but he gleaned as much as he could about them over the years. (JH)

An unidentified UP gas turbine-electric, with second car and fuel tender, appears to make light work of an exceptionally long and heavy load of freight cars, by British standards anyway, which neither then nor now has the infrastructure to manage such tasks. (JH)

engines became even more efficient, the cost issue alone encouraged UP to consider replacing their turbine engines with something more modern. As a result, General Electric were commissioned to design a replacement, with the result being the U50 eight-axled, configured, 5,000hp diesel-electric, twin 2,500hp engined locomotives, the first of twenty-six appearing in 1963 and the last two years later. Inevitably the number of UP 50s in service declined, with the last of them remaining in service until 1969 with two, subsequently, being preserved.

In Britain John Hughes observed all that was happening in Switzerland and the States, collected as much data as he could and continued to study similar developments elsewhere, most notably at Swindon with the GWR. Here, in the late 1930s design engineers were beginning to turn their attention away from steam to consider alternative forms of motive power. Although the GWR had a steam fleet that was probably second to none, thanks to the work and leadership of George Churchward then Charles Collett, as Chief Mechanical Engineer, some believed that this technology had been taken as far as it could go. As any good commercial organisation must do, the GWR began looking to the future, perhaps encouraged by the LMS's development work with diesels and the LNER and Southern Railway's ventures into electrification.

In 1933 and 1936, the GWR had ventured into the world of diesels when purchasing two shunting engines – an 0-4-0 built by Fowler and an 0-6-0 built by Hawthorn, Leslie and Co – for use around Swindon Works. And from 1933 to 1942, they had invested heavily in the development of thirty-eight diesel railcars for use on low density routes. Although a significant step forward these shunters and railcars could hardly be considered replacements for the GWR's massive steam fleet, though did give the motive power department much needed experience of running diesels alongside steam.

In 1941 Frederick Hawksworth, who had been Collett's Chief Draughtsman then Principal Assistant, became CME and here he remained until retiring in 1949. It seems possible that he may have been the driving force behind the company's early diesel programme, or at least gave it his support allowing procurement to continue. Nevertheless, steam remained central to his and the company's plans with a new Pacific engine being of importance to him. It seems that the coming of war put paid to this scheme and there has been some conjecture that this, coupled to a desire to look longer term beyond the end of steam, led Hawksworth and his team to seriously consider alternative forms of motive power. True or not, it seems likely that, at this stage, the gas turbine entered their vocabulary, perhaps encouraged by Stanier's work with *Turbomotive* and Brown-Boveri's experimental engine for the Swiss Federal Railway. If so, the GWR seems to have been quick to consider how this technology might be applied to company needs.

How Hawksworth took this forward is difficult to establish and a timeline even more so. If work started before the war, it would have been natural to approach Brown-Boveri for advice, but being based in the Swiss city of Baden bei Zurich this would have been almost impossible post-1939. So it is more likely that Metropolitan-Vickers, where Henry Guy was still casting his spell over turbine development,

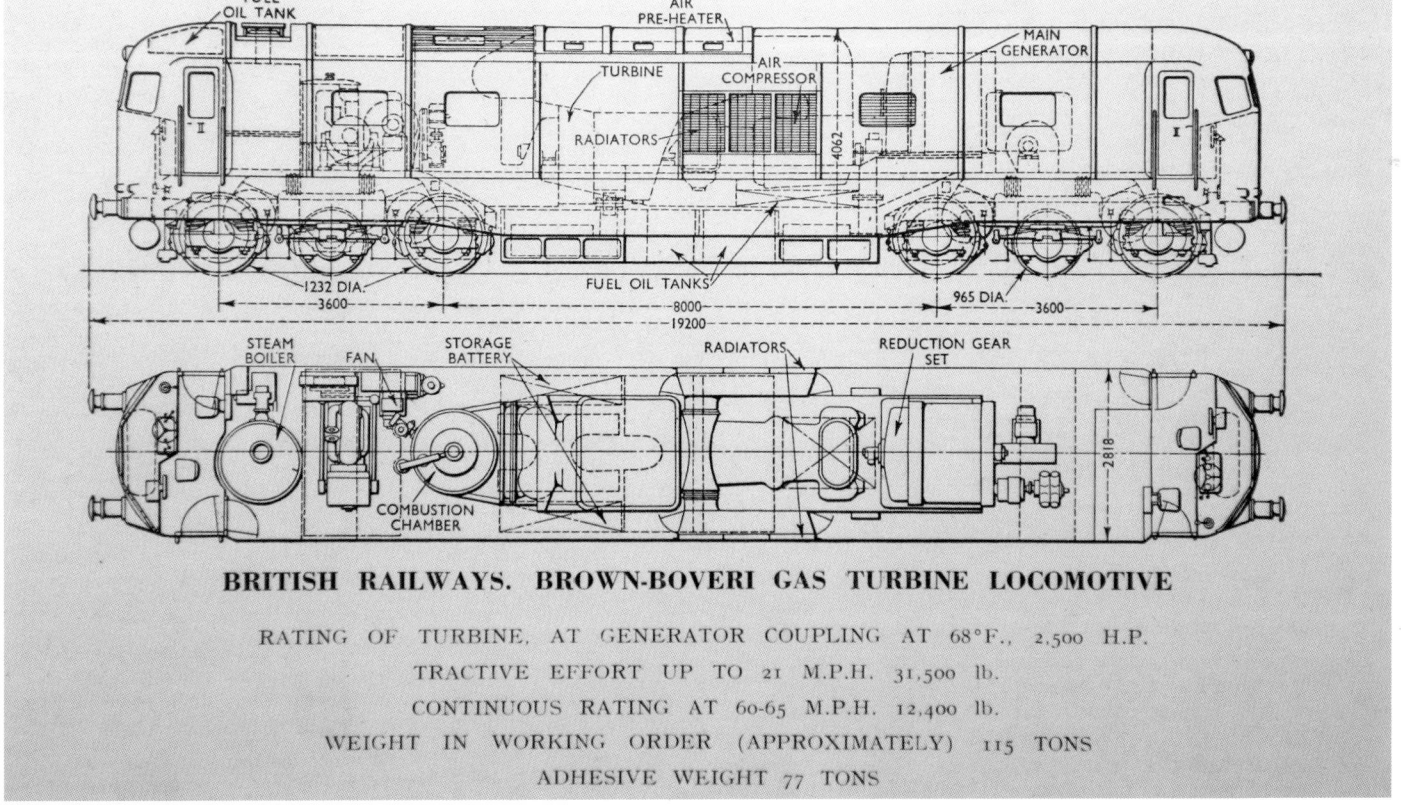

The engine diagram produced before construction of No.18000 in 1947 had begun to show the general dimensions., power outputs and so on. (JH)

was the first port of call instead. With his recent experience of the LMS's turbine Pacific still fresh in the mind, such a move would have been understandable. If so, it is difficult to establish how both sides proceeded or the dates this took place. It has been suggested that as early as 1940 Metrovick may have been commissioned to build a gas turbine-electric locomotive for the GWR. But with the war absorbing so much of industry's time and energy the project was shelved for the duration, it having no obvious application for war use.

When hostilities ended, and the recession that followed eased, it would be revived, but before this happened a second turbine project was initiated. With Switzerland now more accessible, Brown-Boveri were approached, and with their sole gas turbine locomotive now successfully established they had much to offer. In June 1946, Hawksworth, accompanied by Sir James Milne, the GWR's General Manager, visited Switzerland as invited guests of the International Railway Congress being held there. This gave them the opportunity of touring the Brown-Boveri Works to see for themselves what was on offer as regards turbines. They were suitably impressed and later wrote:

> The opinion was formed that not only have the individual problems concerned with each component in a gas turbine-electric locomotive been closely studied, but also the relative proportions of these components in respect of each other been the subject of close investigation, with the result that the complete unit is much more than a collection of separate components mounted together on a main frame.

Clearly struck by all they had seen, they returned home and quickly took action to gain Board approval to open negotiations with Brown Boveri 'regarding the provision of a gas turbine locomotive'. And in September it was reported that 'Brown-Boveri have submitted a tender that provides for delivery in 24 months from receipt of an order … the Minister of Transport has approved the project'.

Such was the progress made during this visit that Hawksworth seems to have obtained approval to acquire a Metrovick designed version as well, according to a report carried in the *Locomotive Magazine* during 1947, which one assumes was endorsed by the GWR's managers before release. However, the Metrovick contract may simply have been a reactivation of an agreement reached in 1939 or '40 and not a newly written agreement.

A second visit to Switzerland by Hawksworth, presumably to confirm that an order had been placed and discuss the project in fine detail, took place in September 1948. However, between his visits in 1946 and 1948 much had changed. First and foremost, the Big Four railway companies, and much more beside, had been consumed by the collective mass of the newly nationalised British Railways. This, of course, would change the dynamics of all new construction programmes, but the two Western Region turbine engines seemed to have passed muster and been allowed to remain in existence. While all this happened, Hawksworth stayed in charge at Swindon, remaining so until the end of 1949, being replaced by another GWR stalwart, Kenneth Cook.

The Brown-Boveri engine was the first to be completed by engineers at the Swiss Locomotive Works in Arlesham and was ready to begin testing locally in November 1949. These lasted for three months and then the engine was shipped to Britain for more extensive trials to be carried out in the immediate aftermath of Hawksworth's departure.

This engine, which was given the number 18000, and the designation GT1, had an A1A-A1A wheel configuration and was fitted with a gas turbine capable of generating 2,500hp. This compared favourably with the GWR's King Class locomotives long established 2,500hp rating. However, the turbine engine could only produce a tractive effort of 31,500lbf to the King's 40,000lbf. As such, the turbine had a maximum speed of 90mph and weighed 115 tons, considerably less than a King and had better route availability. In terms of appearance, it was decided to paint No. 18000 in BR black livery, a silver stripe right around the middle of its body and silver numbers.

In terms of its internal workings, the locomotive was described as a turboshaft type, though differed from a free-turbine turboshaft engines in so far as it had a single turbine to drive both the compressor and the output shaft. The emphasis was on fuel economy so it had a heat exchanger (to recover waste heat from the exhaust) and was designed to run

Two photographs of No. 18000 thought to have been taken in Switzerland after its initial trials had been completed and the engine had been painted in BR's selected colour scheme ready to be shipped to Britain. (JH)

on cheap heavy fuel oil (it was also able to burn light oil but this was intended only for start-up purposes). After passing through the heat exchanger, the pre-heated air entered a large combustion chamber where the fuel was injected and burned.

Sadly, it proved to be a troublesome machine in service. Ash from the heavy fuel oil damaged the turbine blades, and the combustion chamber liner required frequent replacement due to damage. The electrical control systems were extremely complex for the time and gave much trouble, maintenance of the electrical equipment in a steam shed with all its dirt and grime making things worse. Then, part way through its life, one of the traction motors failed and instead of repairing or replacing the damaged parts, it was simply removed, leaving the locomotive with only three traction motors, so preventing it from achieving its full power output. And then, it suffered more severe damage when its heat exchanger was destroyed in a fire at Temple Meads, caused by combustion deposits in the exhaust side of the heat exchanger igniting.

The locomotive was also found to be much more expensive to run than expected. The efficiency of a gas turbine reduces dramatically at low power outputs so, to achieve respectable fuel economy, a gas turbine locomotive needs to be operated at full power as much as possible, with periods of part-load running reduced to a minimum. However, it turned out that, even on demanding express passenger

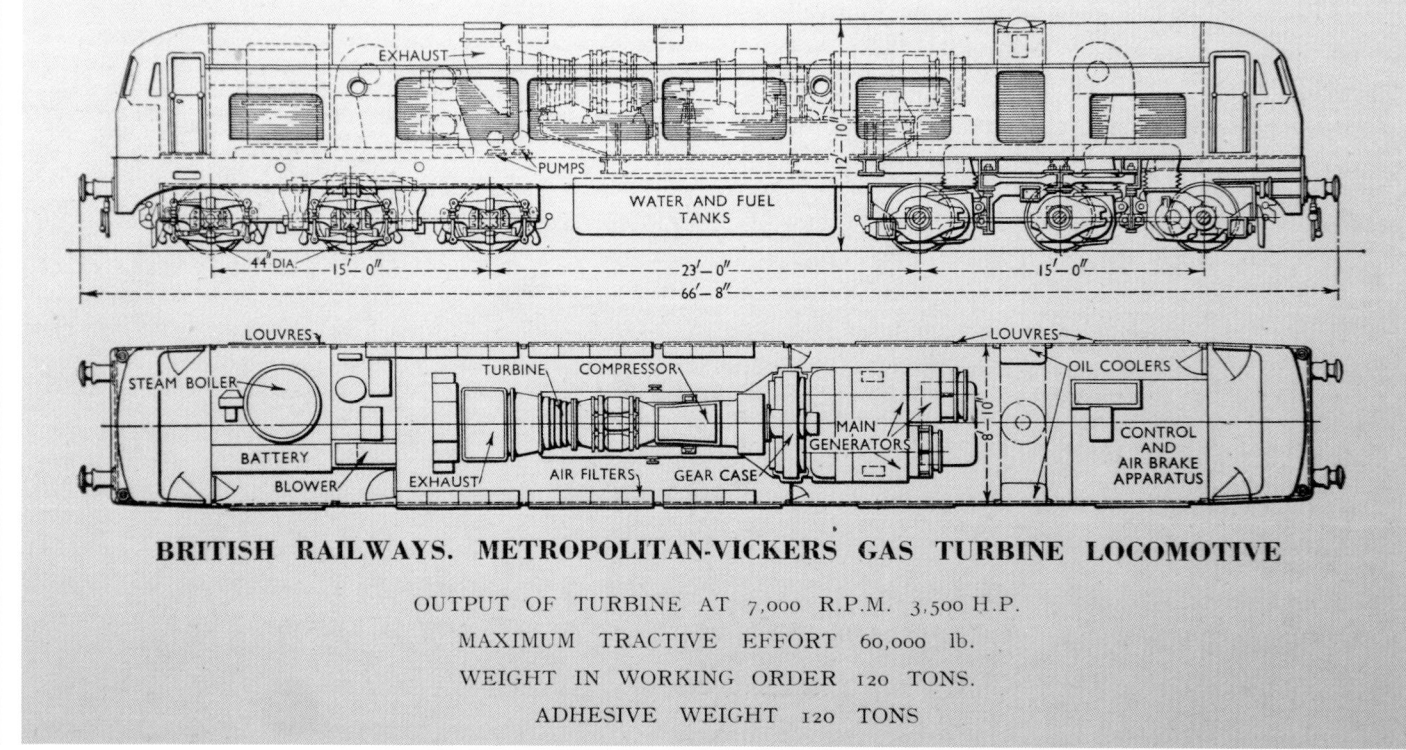

GTEL No.18000 photographed at Swindon in April 1950 when under test, on this occasion with the old GWR dynamometer car and thirteen carriages attached all weighing some 436 tons. During this exercise, the train would travel to Bath, Bristol and Badminton before returning to Swindon. In due course, the results would be published in report W7, a copy of which John Hughes acquired. (Author)

No. 18100's diagram as kept amongst John Hughes's papers. He seems to have shown as much interest in this engine and No. 18000 as he did *Turbomotive* and the GTEL projects in Switzerland and the USA. (JH)

schedules, it was not possible to operate the locomotive on full power for very much of the time and extended periods of part-load operation were inevitable, which resulted in even heavier fuel consumption.

While No. 18000 was being put through her paces the second of the gas turbine-electric projects was slowly coming to fruition at Metrovicks Trafford Park Works in Manchester. In this case a Co-Co wheel arrangement was employed with the turbine rated to produce 3,000 horsepower. The engine weighed 129.5 tons and its tractive effort was predicted to be somewhere in the region of 60,000lbf and, like No. 18000, it was hoped that it would be capable of a maximum speed of 90 mph. And to match its Brown-Boveri sister she was to be painted in BR black with identical silver markings around her body.

The new engine shared turboshaft technology with No. 18000, with only one turbine to drive the compressor and the output shaft, but there was no auxiliary diesel engine and the turbine was started by battery power, with the main generators being used as a starter motor. Interestingly, the engine's specification placed greater emphasis on power output than on economy, so in service, perhaps inevitably, its fuel consumption proved to be high. The engine was also designed to use jet turbine fuel, but this was more expensive so making the engine less economic to run than No. 18000, which used more cheaply produced heavy fuel oil. There was also the question of safety to consider, aviation fuel being more toxic and so more difficult to handle in day to day service. It also had a much lower flashpoint than heavy fuel oil and was more susceptible to the risk of water contamination.

Judging by the papers he kept, the Metrovick turbine engine seems to have been of more interest to John Hughes than No. 18000. In 1951, when No. 18100 (which became GT2) appeared, BR's Railway Executive produced an internal 24-page booklet describing the engine in some detail. Hughes obtained a copy by some means and carefully highlighted the design issues he presumably thought most important. For the gas turbine these included:

The prime mover is a single open cycle gas turbine without heat exchanger and the cycle of compression, heating and expansion of the air is carried out in a compression, combustion chamber and turbine arranged in line and built into a straight through unit.

The compressor is a 15 stage axial flow with a pressure ratio of 5.25 at 7,000rpm and a mass flow of 50lbs per second… The turbine is a five stage unit running in two sleeve type bearings…The compressor and turbine cylinders are connected together by a tubular member surrounding their shaft coupling so that they constitute a single structural unit.

In the construction of the compressor the cylinder is made of malleable iron and the rotor is a forged steel drum. The moving blades, which are machined from stainless steel bar and fixed blades rolled from similar material, are retained

No. 18100 under construction at the Metrovick Works in Manchester. In this view we can see the compressor (nearest the camera) and the rear of the main generator. (JH)

Two views of gas turbine engine No. 18100 when new and under test. (Below) Seen here at Goring troughs pulling the *Bristolian* service on 24 April 1952 (To the right) The engine is captured at Bristol Temple Meads a few days later, on 31 May 1952, again rostered for the *Bristolian* to Paddington. Unfortunately, it failed to start when required and is seen here about to set off for Swindon. John Hughes would later write that N0. 18100''had a neat, business-like, modern look, but in practice it was too expensive to run encouraging Swindon, when BR's Modernisation Plan of 1955 was implemented, to go down the diesel-hydraulic route followed in Germany with some success'. (JH)

in dovetail slots…The bearings are lined with white metal and lubricated and cooled by a copious supply of oil [to which Hughes has added 'I wonder what oil was used?'].

Turbine construction employs special heat resisting materials, the cylinder being an austenitic steel casting and the rotor an austenitic steel forging. The bearings are of similar design to those of the compressor but additional cooling is provided by a flow of compressed air from an intermediate stage of the compressor.

Fuel combustion assists in the acceleration (after the turbine has been accelerated up to a self-sustaining speed by the main generators acting as motors fed from the starting battery and the driver has actuated the starting button which makes the process entirely automatic) and at about 2,500rpm the battery automatically disconnects from the generators and the turbine continues its acceleration under its own power to 4,000rpm.

Even though Hughes had so recently joined English Electric's Gas Turbine Department, and was still five years away from heading their Turbine Locomotive project as Chief Designer, his interest in the fine detail of the second Western Region project is most revealing. And it didn't stop with the turbine itself but spread, as his notes reveal, to all parts and functions of No. 18100 – its reduction gear, the main and auxiliary generators, the traction motors, the ergonomics of the cabs and the controls and equipment they contained. Then there was the structure of the frames and the running gear and much else beside. Finally, he kept a close eye on the way the engine performed, acquiring various reports, presumably obtained from contacts in the commercial world, Swindon or the Railway Executive. Even though English Electric were still a long way from commissioning such a locomotive themselves, at least one of their number was actively thinking about what this might entail. One wonders whether Hughes had already been given the green light to consider such a proposal and, so, had begun mapping out a programme of work.

For the engineers at Swindon, who had pursued two gas turbine projects passionately for so long, only disappointment awaited them. No. 18000 had proved somewhat unreliable and its fuel consumption was considered too high to be economically viable. Nevertheless, it remained in service until 1960 when withdrawn and placed into store at Swindon, having covered some 350,000 miles. Then in 1964 it was returned to mainland Europe, having been offered to the Office for Research and Development (ORE), based in Utrecht, by BR as a test

During 1953, following a major failure near Newbury on 31 January, No. 18100 was towed to Swindon for repairs to her electrical system. Whilst in the works, the opportunity was taken to modify the bogies and fit stiffening beams. To do this, the engine was stripped down to its frames, in which state she is photographed beside the 1951 built Britannia Class 7 Pacific No.70022 *Tornado*. (Author)

NBL and Co's initial drawing of the proposed pulverised coal powered gas turbine diesel showing the most basic of design details. This drawing was attached to a thirty-page specification that the company produced for the Ministry of Power and Fuel which John Hughes acquired. Although showing little interest in the fuel used, he was intrigued by its mechanical transmission which judging by his written comments, influenced his thoughts when the time came to design GT3. (JH)

vehicle to help study the problem of adhesion between steel wheel and rail in traction. To do this, No.18000 was much modified and her gas turbine removed, then for ten years or more undertook test work before being withdrawn from service again. In 1975 she entered preservation, first in Austria and then in Britain, where she resides to this day.

No. 18100 was not so lucky. Prolonged testing revealed flaws in the design, most noticeably in its high rate of fuel consumption, which various modifications failed to correct. As a result, the experiment came to an end and she was withdrawn in 1958 then placed in store at Swindon. In due course, Metropolitan Vickers negotiated her return to allow engineers to conduct an experiment by converting her into a prototype 25 kV AC electric locomotive. In this form, and renumbered E2001, she underwent further testing until withdrawn for a second time. However, on this occasion there would be no saviour and she was scrapped in late 1972, so bringing a useful, but ultimately unsuccessful experiment to an end.

Intriguingly, there was one other form of turbine Hughes studied closely at this time – a coal fired version. In the 1940s and 1950s research was conducted, in both the US and UK, aimed at building gas turbine locomotives that could run on pulverized coal. With war generating a very heavy demand for oil, the Allied powers were soon struggling to meet all their needs in the face of strict rationing. This was especially so in Britain. This continued when the war was over and so alternatives were considered and eagerly sought as the railways sought to modernise their fleets. Coal, already the mainstay of steam locomotion, was one option contemplated and being available in huge quantities made its use, in some derived form, seem most attractive.

Quite simply pulverisation takes place when coal is ground to a fine dust and mixed with air blown into a combustion chamber,

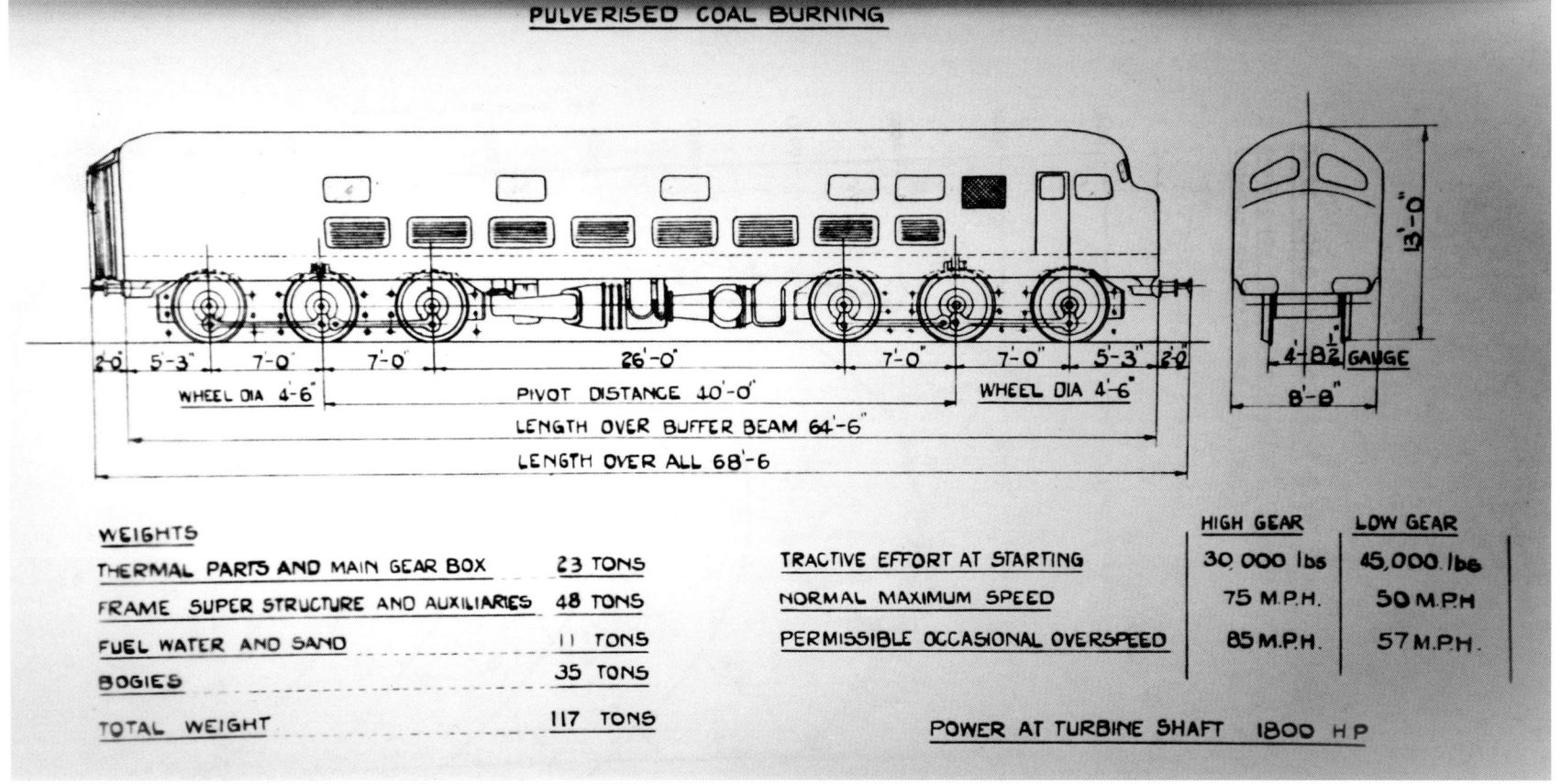

where it is burnt to create a high degree of thermal energy. The only problem, as with untreated coal, is that this creates dust and ash that can be damaging to machinery. Nevertheless., experience had shown that pulverised coal was an ideal material for use in driving large turbines that lay static within power stations and where the contamination problem could be carefully monitored and controlled.

When it came to be used in mobile, high tech engines, the extent of this problem soon became apparent. Here the erosion of turbine blades by particles of ash entering the machinery created a seemingly unresolvable dilemma. This was highlighted most vividly during in 1946 by Northrop-Hendy where engineers attempted to adapt a Northrop Turbodyne aircraft engine for use in a locomotive, using pulverised coal rather than kerosene as its source of power. To help them in this endeavour Union Pacific bequeathed their 1936 built M-10002 streamliner locomotive to Northrop-Hendy, late that year, so they might develop the concept further. By the end of 1947 it was clear that success was eluding them and the project was cancelled and the General Electric developed gas turbine engine adopted instead.

This research was noted in Britain and despite its failure the UK's Ministry of Fuel and Power placed an order, during December 1952, for a coal-fired gas turbine locomotive for use on the nationalised network by BR. One can only think that the cost of oil then and the availability of cheap coal in the UK drove the powers that be to seek such a solution. If successful, it could have had many obvious benefits, which a paper produced in May 1951 by the Ministry of Fuel and Power described as 'potentially reducing reliance on expensive, imported oil from a volatile region and, if successful, will give BR a chance to modernise its aging, steam driven locomotive fleet without recourse to oil fired solutions'. After careful consideration the North British Locomotive Company were selected to build a prototype with a suitable turbine being supplied by C.A. Parsons and Company.

Here John Hughes picks up the story, having been given a copy of a Government sponsored paper

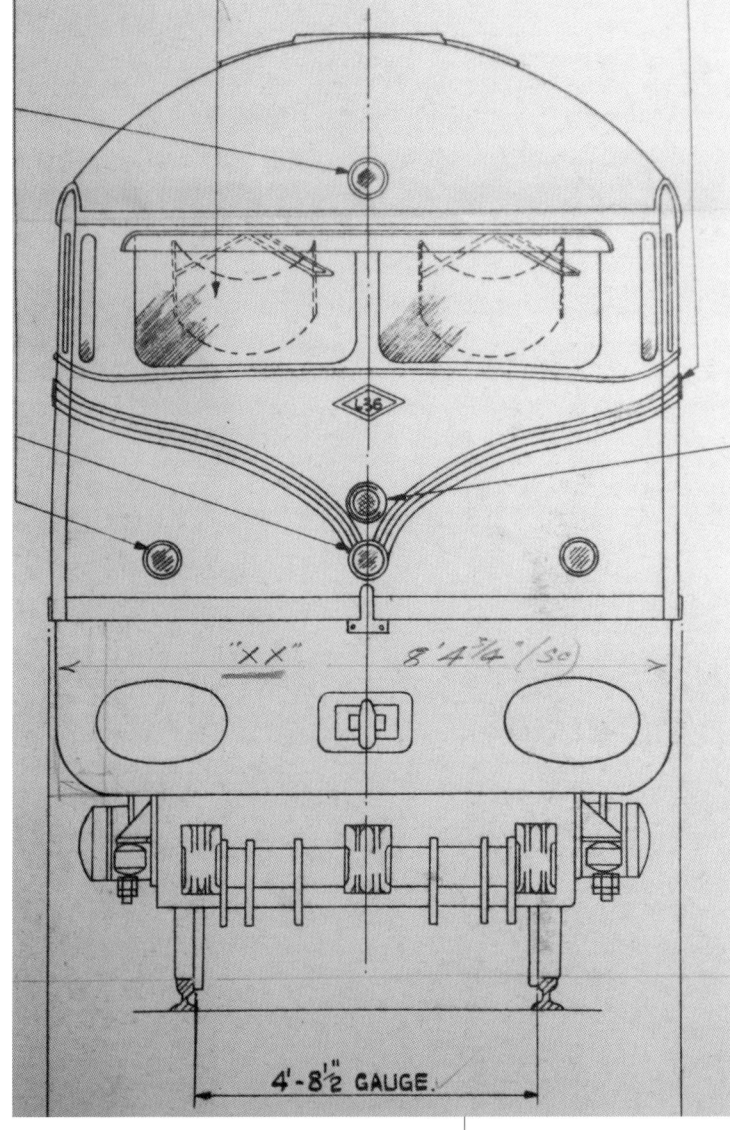

The pulverised coal powered gas turbine mechanical engine takes shape. Here again Hughes managed to acquire this and many other later stage drawings prepared by NBL and C.A. Parsons and Company. (JH)

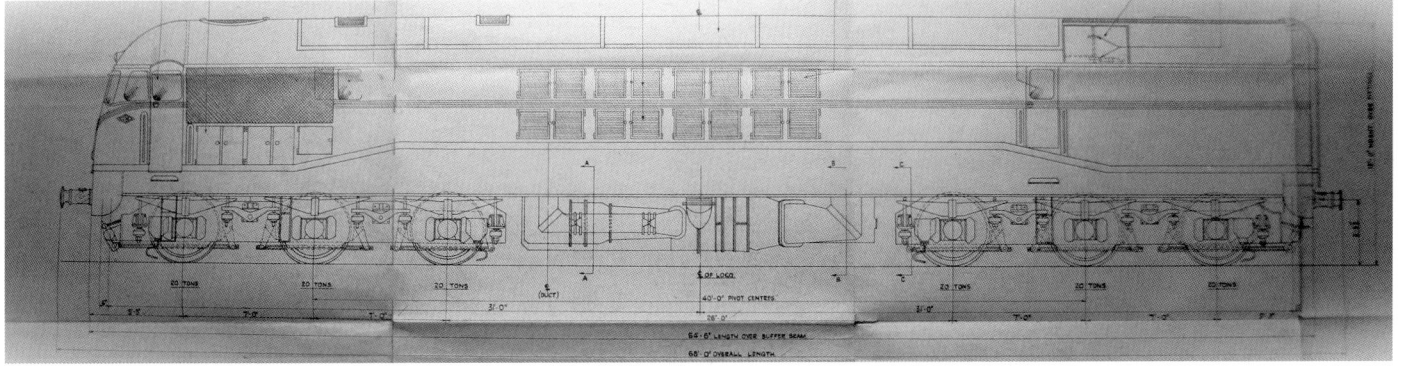

published on 8 March 1951. Once again it is interesting to follow the particular areas he has highlighted in the general specification:

Leading Particulars

Power at turbine shaft	1800hp.
Max rail hp	1600hp.
Wheelbase	0-6-6-0.
Length overall	68ft 6in.
Total weight	117 tons.
Maximum axle load	19 ½ tons.
Normal maximum speed	75mph.
Starting tractive effort	30,000lbs.
Maximum TE at low speed	45,000lbs.
Fuel capacity	6 tons (or potentially 500 miles range at average to maximum speeds).
Maximum turbine temperature	1300 degrees F.
Maximum speed power output	8,000rpm.

He then moved on to highlight more specific issues, particularly the proposed mechanical transmission:

> With a power at the rail of 1600 hp the locomotive should be able to handle fast passenger trains of more than 12 coaches and freight trains of 620 tons or more. A two range gearbox will be required. This will provide a choice of gears suitable for either type of work. When geared for tractive effort at starting at 30,000lbs the top speed in normal running is 75mph but occasional over speed to 85mph will be allowable.
>
> The gas turbine itself will be of a robust design and involve no new principles. It will be slung below the frame in an accessible position. The compressor will be driven by a high pressure turbine from which the gas passes to the low pressure turbine on a separate shaft. The torque characteristics of the power turbine resemble those of the conventional steam locomotive permitting direct mechanical drive. There is a close analogy with the LMS steam turbine locomotive number 6202 built before the war and is still in service today.
>
> The power turbine will drive the main reduction gearbox to permit a choice of two speed ranges forward and one reverse with a differential which divides the output between the two shafts running fore and aft.
>
> The main difficulty in achieving a satisfactory locomotive arrangement has been the size of the heater and the limits placed on re-arrangement by the need to consider the flow of gas. Early attempts at an eight-coupled rigid wheel base showed great promise but the overall length was too great unless the fuel was carried in a separate tender and with it the train heating boiler, which would then have to be fired separately. Even here a choice had to be made between accepting either limited visibility or an automatic control for a train heating boiler placed out of reach of the crew
>
> It was hoped to provide two cabs and a locomotive that would not need to be turned…. However, the arrangement chosen has been kept simple as it was thought that unnecessary duplication would merely complicate the initial development work without providing the answers to any new problems.
>
> The work turbine will be connected by a quill shaft to the main gearbox in which the speed of the turbine shaft (8,000 rpm max) is reduced to a speed suitable for the bevel drive in the bogie gearboxes (2,500 rpm)….The gears must be in constant mesh and be connected individually to the shaft by engaging the respective dog clutches. The gearbox output shaft should drive a differential, in order to distribute power equally between the bogies…. The bogie gearboxes must be of identical design and consist of three components. The first to contain the double reduction gears and universal joint, the second has the final drive and the third contains the quill shaft and flexible discs to overcome the relative movement between the output shaft of the second reduction gear and the input shaft of the final drive."

Having consumed this 30-page report, Hughes then penned a short memo, with no addresses listed,

A Slow Burning Revolution • 63

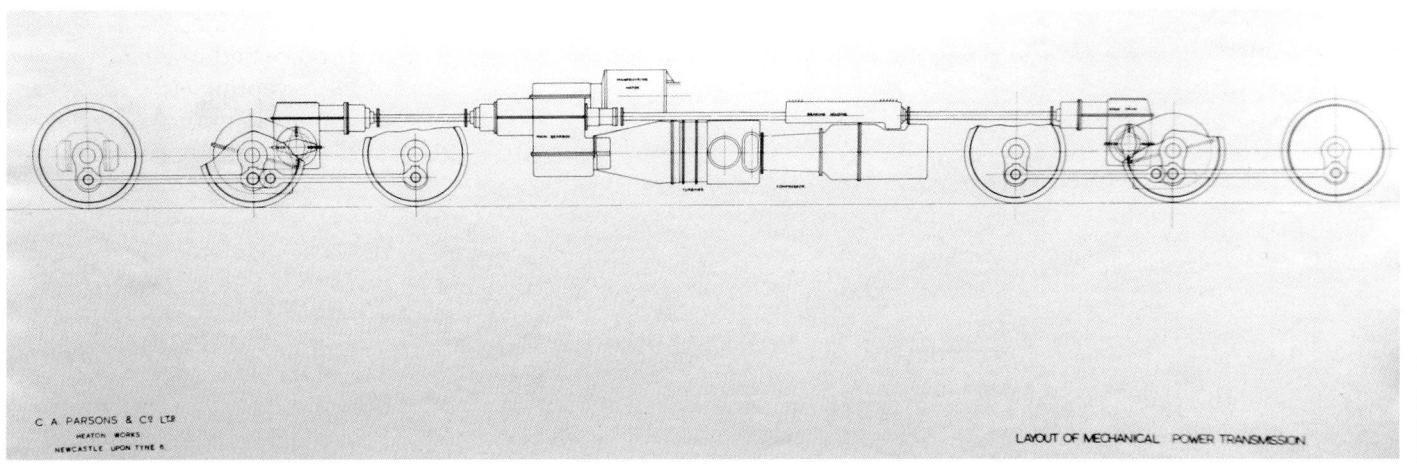

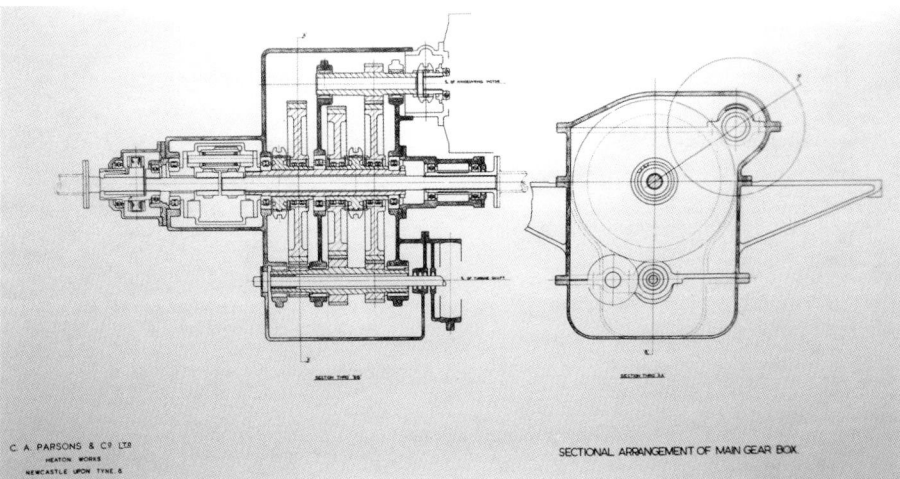

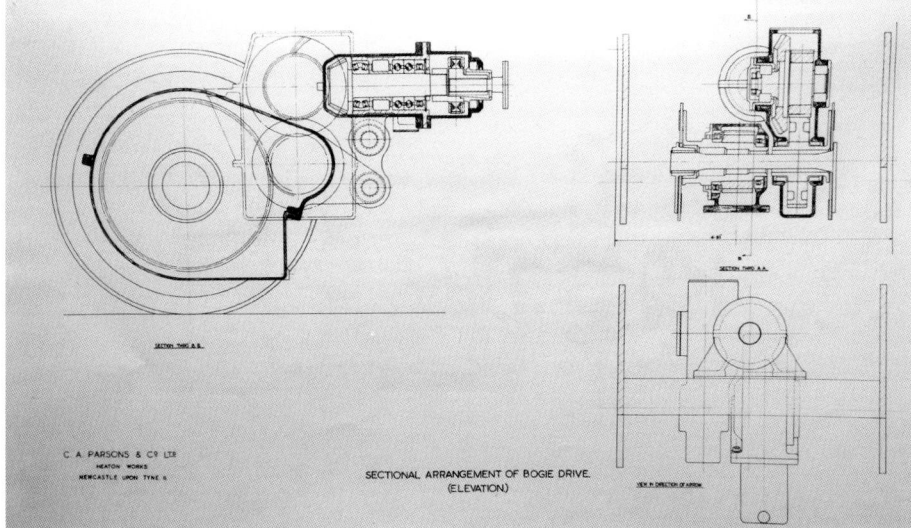

When designing the pulverised coal powered turbine locomotive its specification included many drawings. It is interesting to note that of all these John Hughes selected these three to include in his design papers for GT3 suggesting that the NBL's proposals for a mechanical transmission were of interest to him more than any other aspect of the design. (JH)

which he then attached to the front cover. In it he wrote:

> There is much to consider here, particularly in the application of a mechanical transmission, which seems to me to be a simpler, more practical solution than a gas turbine-electric type engine (and probably cost a great deal less to construct). A single cab also has many benefits, as has the application of known turbine technology without new embellishments. Here, our now proven EM-27, if built to a more compact specification, might be employed to good effect.
>
> Clearly the pulverised coal concept, as the Americans have already discovered, though driven by the reality of high fuel oil costs, is a non-starter. Steam engines are better able to stand the side-effects of ash and dust pollutants, a turbine far less so. In the event I will be most surprised if the Ministry of Fuel and Power's proposal will succeed or even get beyond the drawing board.

Hughes then withdrew five of the many drawings contained in this paper, put them to one side and then marked them as 'being of great use if and when English Electric are commissioned to build a gas turbine engine'. In 1952, when he wrote these words, the chances of this happening must have seemed slim, BR being very heavily committed to build nearly 1,000 new standard class steam locomotives. And during the year even *Turbomotive* was withdrawn and converted into a traditional steam engine. Meanwhile, dieselisation and electrification programmes seemed to have stalled despite the best efforts of some, including English Electric, to kick start a much-needed modernisation programme. But coal was still king, as far as British Railways were concerned, and, even when obsolescence beckoned, had to be played out to an unsatisfactory conclusion.

Meanwhile, John Hughes dreamed his dreams of building a successful gas turbine engine, continued to gather and sift the results of nearly half a century of experiments and planned for a future he may have thought might never arrive. Then, in 1955 two things changed, placing him closer to his goal. First of all, he was promoted to become Chief Designer of English Electric's Gas Turbine Project Group and given a brief to design and develop a locomotive. At virtually the same time, BR finally grasped the need to look beyond steam, to plan its demise and begin a true period of modernisation. BR might not accept the need for gas turbine engines, but any company willing to develop the concept further at least stood a better chance of success than before. It was a thought that obviously occurred to Hughes, who late that year presented a paper to his Board that concluded with the words, 'The development of the gas turbine locomotive to a total adhesion bogie type with mechanical transmission can be foreseen along clearly defined lines and with success will lead to the eventual disappearance of the diesel engine from a wide field of rail traction.'

It was an ambitious statement, made by a man in thrall to his science, but he faced many challenges and huge barriers if he were to succeed. But he was a determined, committed man who believed in the turbine's future and wouldn't easily be baulked by any opposition.

Colour Section
GT3 – A DESIGN HISTORY IN COLOUR

From early in his life as an engineer John Hughes seems to have been fascinated by the potential contained in turbine engines in their various guises. It was a fascination he shared with a number of railway engineers, whose work Hughes avidly followed. This was none more so than in the case of William Stanier and his work in developing his turbine powered steam 4-6-2 locomotive *'Turbomotive'*. Although Hughes' career took him into aviation and motoring areas, when the opportunity arose he applied his accumulated knowledge to the design of railway locomotives. And so, GT3 was born.

The greatly anticipated public launch of Stanier's prototype turbine engine, based on his 4-6-2 Princess Class locomotives, in 1935. It proved to be a successful experiment but despite Stanier's advocacy did not lead to the wider development programme he hoped to foster. 6202/46202 remained a singleton engine until rebuilt into a more conventional form in 1952. Hughes, who saw this engine operating on many occasions was, he reported, profoundly influenced by this project and its creator. (TC)

Above and below: During the 1940s designers in the USA were seized by the turbine concept and conceived ever more elaborate plans to exploit its potential. Hughes followed this work, corresponded with these engineers and accumulated many brochures such as this from Baldwins of Philadelphia. Three of these M1 steam turbine electric locomotives were built in 1947/48 but proved costly to run and had all been scrapped by 1950. (JH)

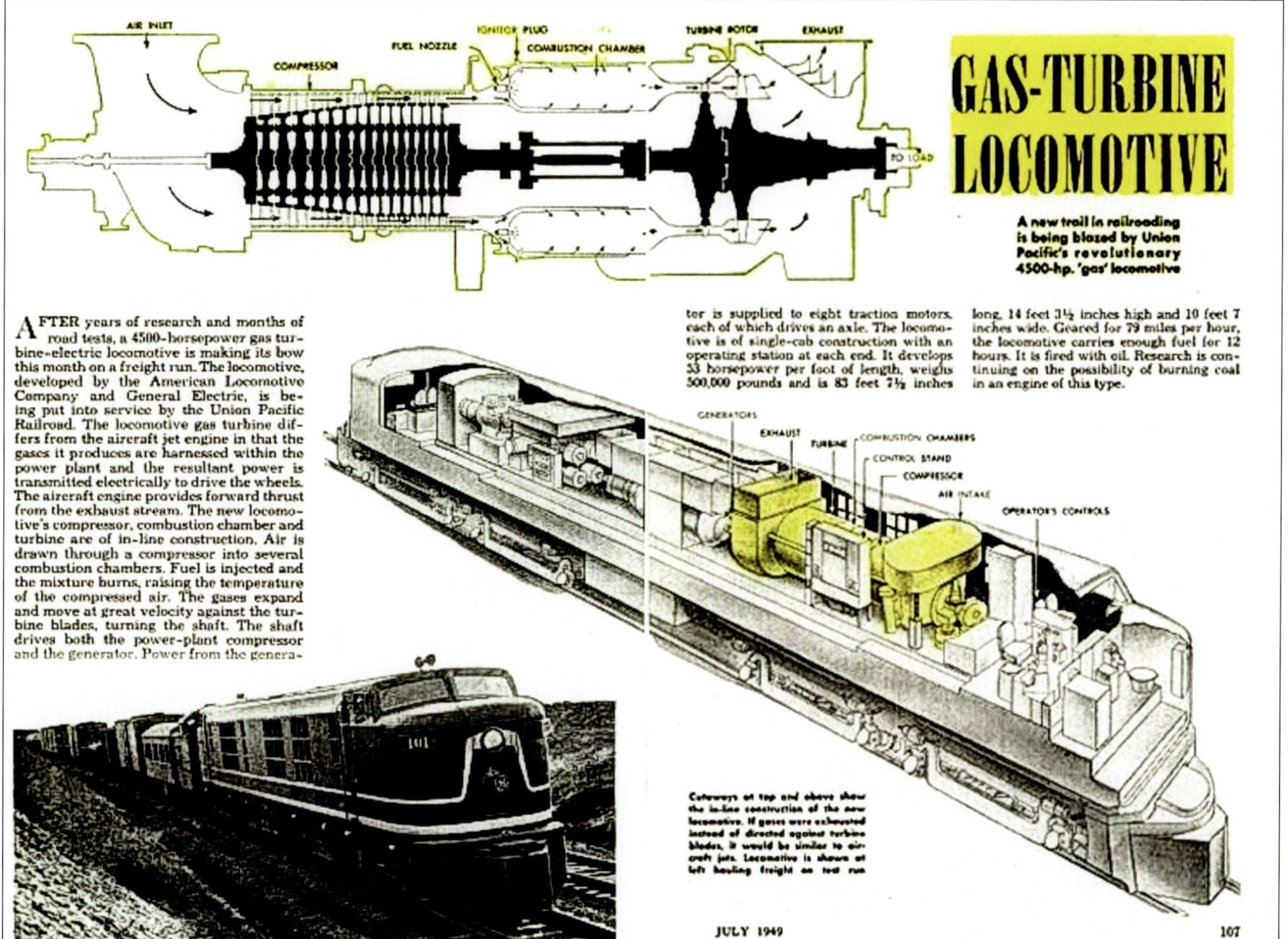

Above and overleaf: In the late 1940s oil fired gas powered turbine locomotives began making their appearance. ALCO produced a prototype in 1948 (above) and this found favour with the Union Pacific who acquired 55 in total for freight work. Over the next twenty years these reliable work horses accounted for, on average 10% of all UP's freight traffic and remained in service until the late 1960s. The number of photographs (example overleaf), reports and brochures acquired by Hughes concerning these engines suggests a deep interest in their design and performance. (JH)

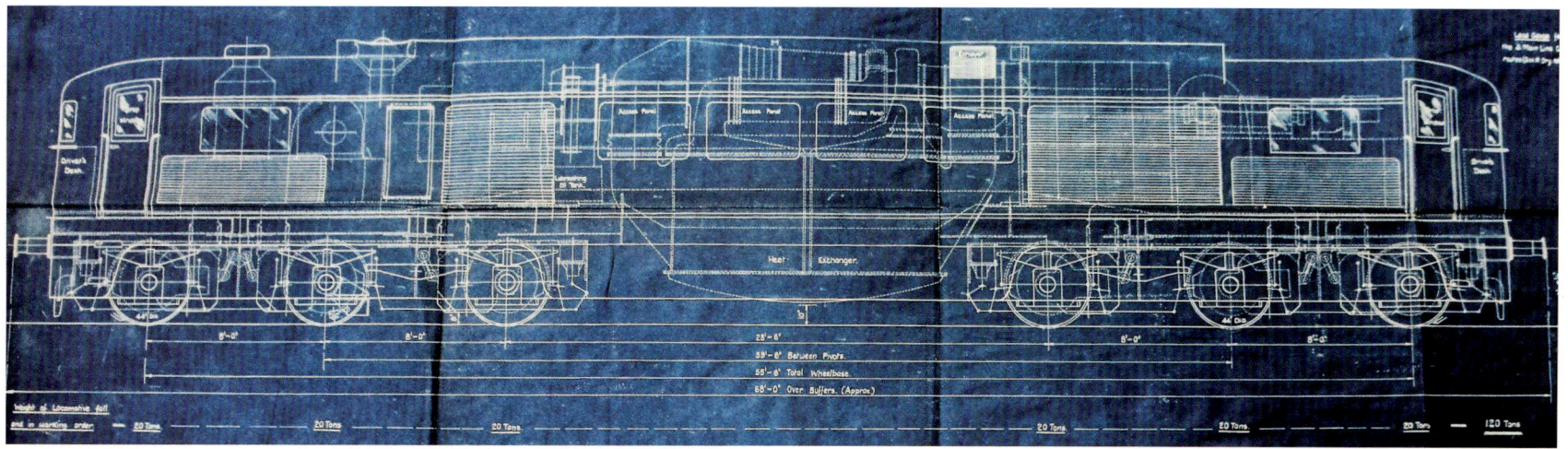

During the war years the GWR began to show some interest in gas turbine electric designs and contracted Metropolitan-Vickers to construction a single Co Co engine (top and left). With the conflict raging the project was placed in abeyance, but in 1951 BR reactivated the work and engine No. 18100 went into service. In the meantime, a second such engine was built by Brown Boveri (above right) for BR's Western Region, to an AIA-AIA configuration and this appeared in 1948. Neither project led to more engines of the type being built, though the turbine experiments undertaken clearly influenced John Hughes who collected a great deal of information about both types (Author).

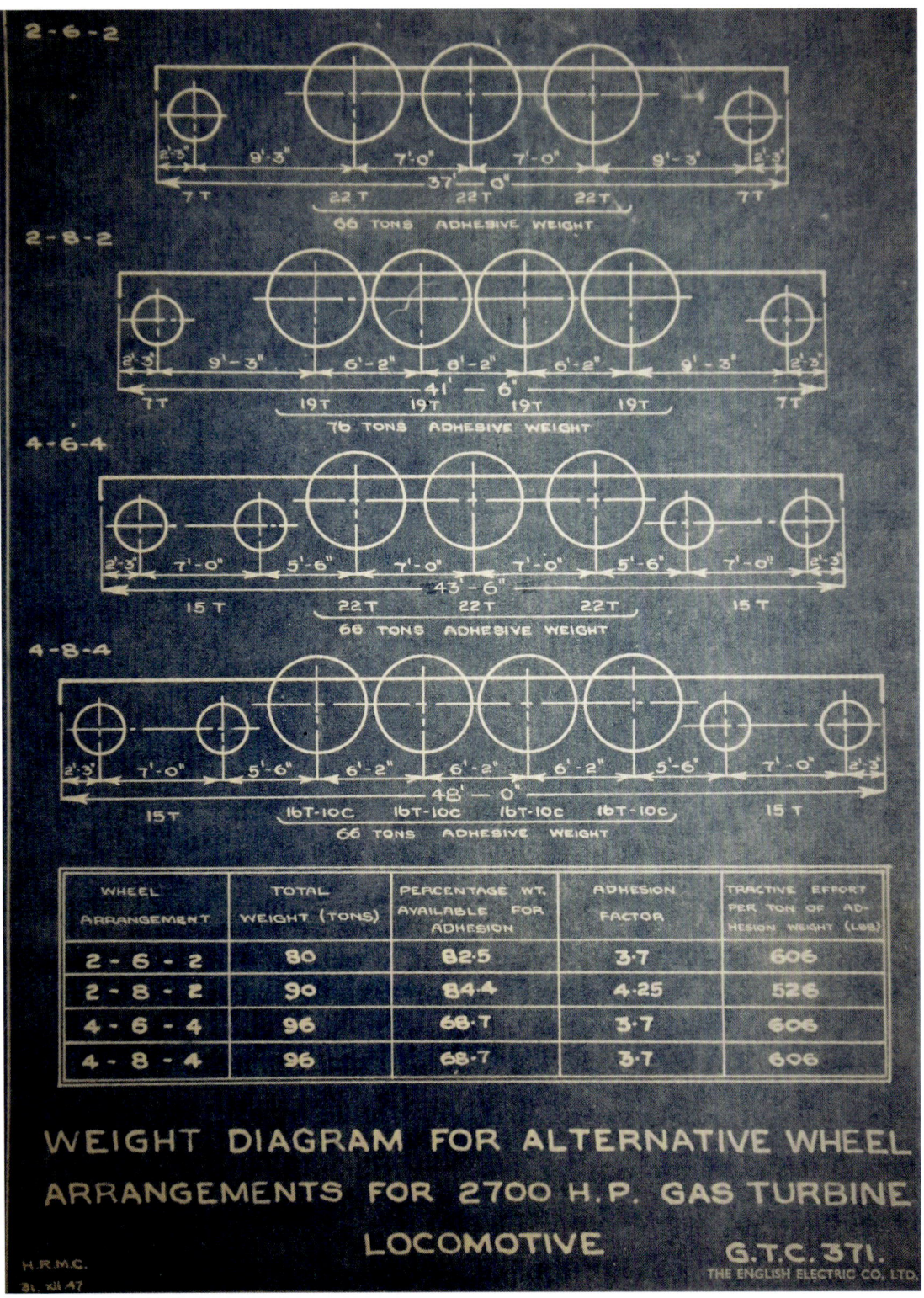

Left: **As Brown Boveri** and Meto-Vickers completed design work on their turbine engines, English Electric were considering their own version, having gained wide experience of steam turbines in industry. There were several concept papers produced in this early exploration, all of which Hughes retained. A great deal of discussion seems to have focussed on the potential of each wheel configuration, as demonstrated in this 1947 produced chart. As GT3 developed the company expressed a preference for a 4-6-0 design and this solution was adopted. (JH)

Opposite above and opposite below: **During 1947/48 Hughes**, aided by Bill Allen, a friend and graphic designer, began considering how the English Electric turbine engine might appear if built with different wheel configurations. Judging by their correspondence and the many drawings that were produced, it was a process which both men enjoyed considerably in advance of the more complex design work starting, these are just two of the drawings that were produced during this period of 'blue-sky' thinking. (JH/WA)

Colour Section • 71

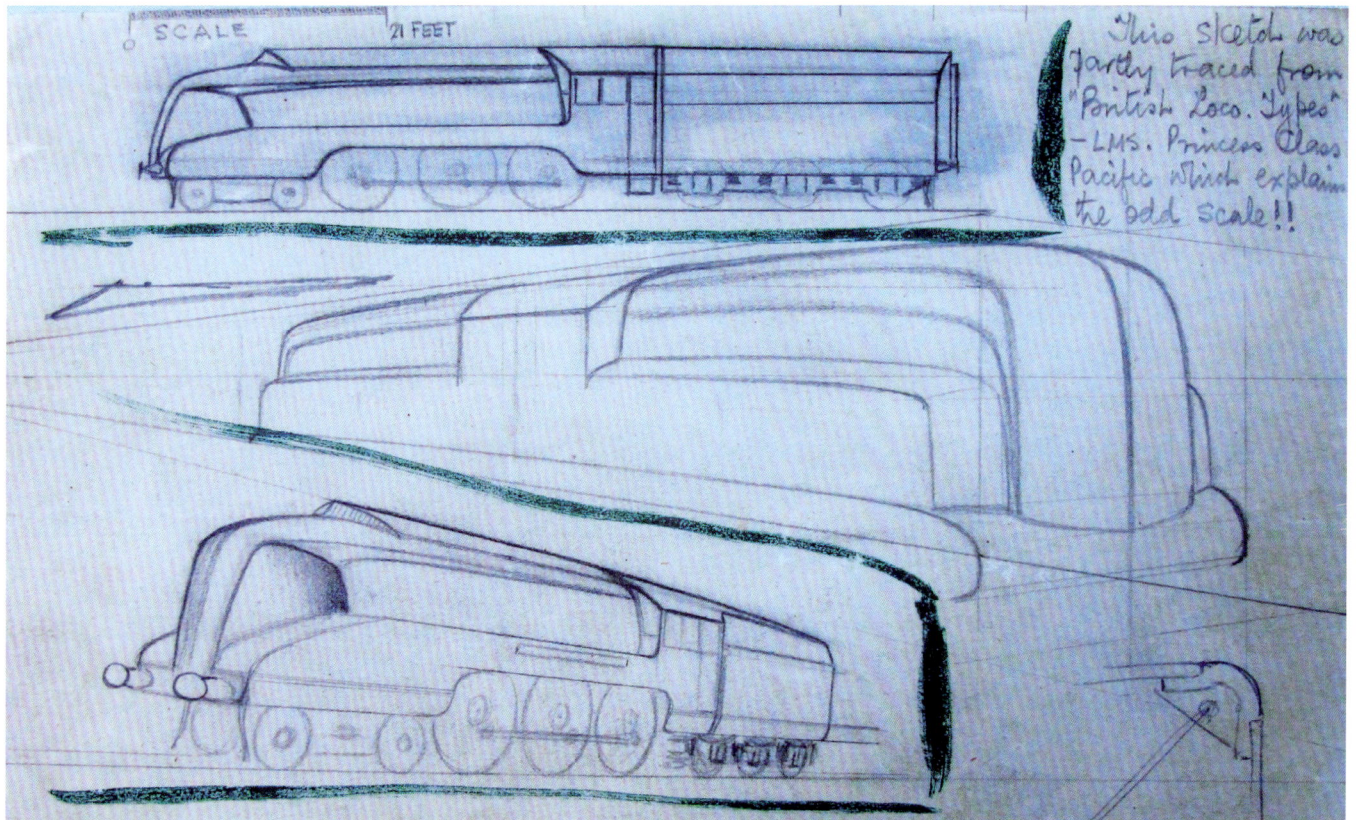

Right: **GT3 was**, by any standards, an unhurried project, work on diesel and electric locomotives by English Electric taking priority, but, by degrees, Hughes and his teams brought their plans to fruition, of which this diagram is just one simplistic example. (Author)

Below: **To demonstrate** how GT3 would work the English Electric team produced a brochure containing a number of coloured drawings of which this is one example. These were circulated amongst potential customers, including BR. Such was the interest expressed that EE continued with the project. (Author)

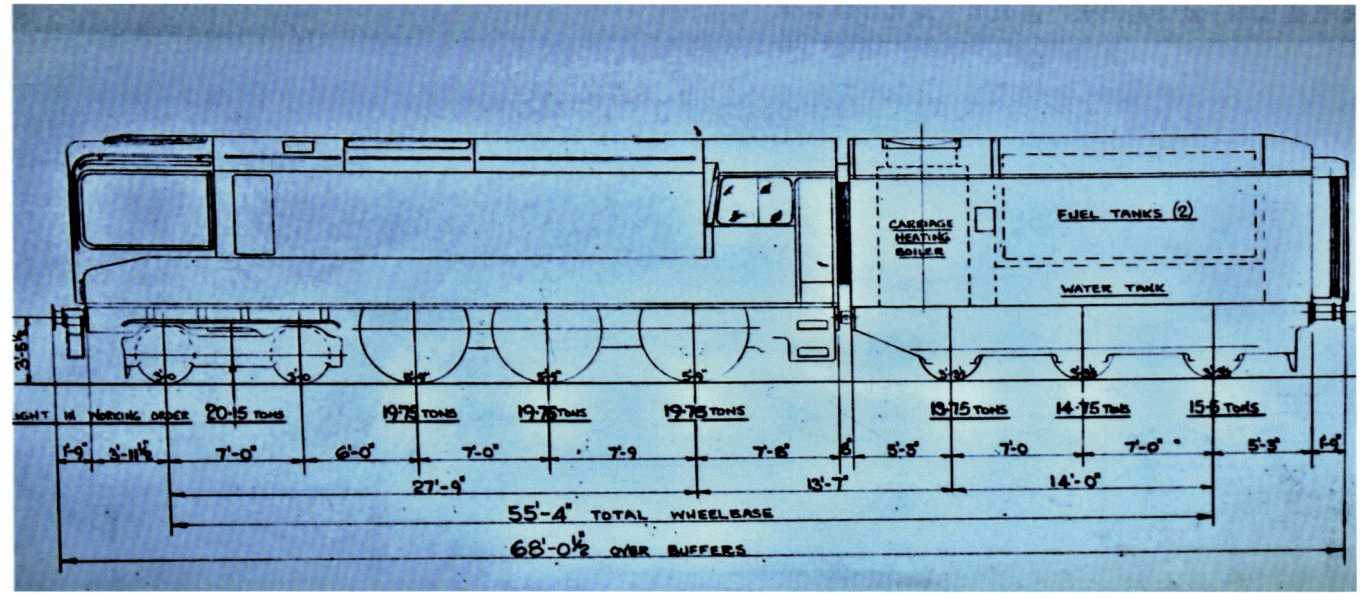

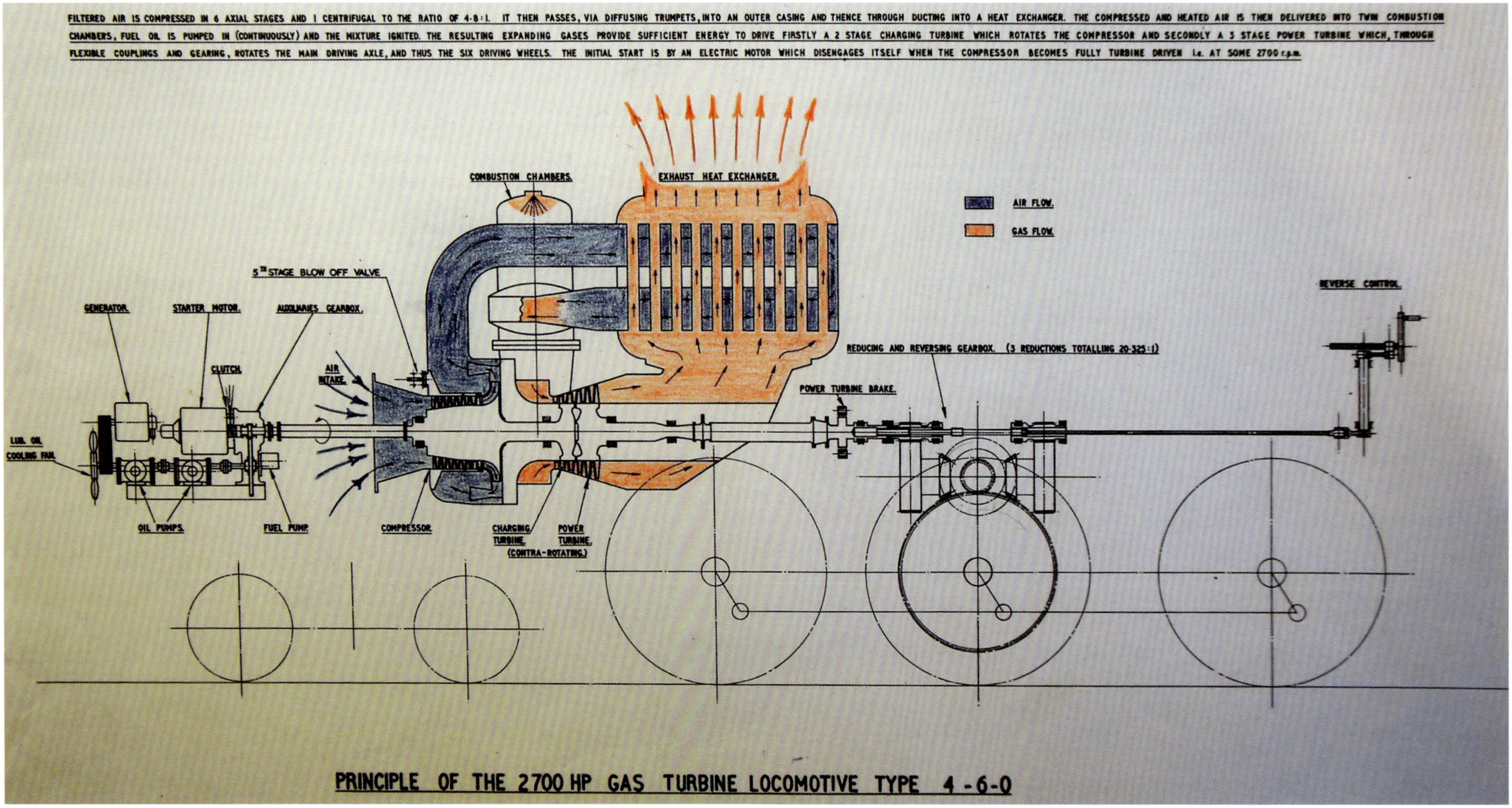

As the design progressed Hughes and Allen contemplated colour schemes for the prototype locomotive and a series of paintings were produced for English Electric to consider. Each drawing also gives some idea of the way the exterior design was heading – around the cab and the engine's front end in particular. However, the engine is recognisably GT3 as she would appear in 1959 when completed at the Vulcan Foundry Works at Newton-le-Willows in Lancashire. (JH/WA)

Left and overleaf spread: **When still** incomplete GT3 underwent tests with various loads over varying gradients, a process designed to highlight faults and suggest modifications. This proved to be the case and, amongst other things, the lubricating system, then the turbine blades which were found to be fouling the stator casing, underwent redesign and rebuilding. A period of trials running was then undertaken at the Rugby Test Centre in 1957/58, before being returned to EE for completion and the fitting of her body. Its unusual shape and distinctive brown colour scheme made her stand out from the crowd and drew much press attention as a result. Over the following months she became a common sight over parts of the network as efforts were made to prove the design. These photos capture her during this period. (JH/Author)

Above and opposite: Good PR is essential to selling any product and this was no different for GT3 and over two years the locomotive became a frequent presence in newspapers and journals; her non-standard appearance making her an object of curiosity if nothing else. These are just three examples of the many items that appeared. (Author)

ENGLISH ELECTRIC G.T.-3
This locomotive was driven by a gas turbine or jet engine of 2,700 h.p. However, this interesting loco' was never put into regular service, but was used as an experiment on British Railways during the 1950's.

50th Anniversary Year

THE INSTITUTION OF LOCOMOTIVE ENGINEERS

President:
D. C. BROWN, C.B.E.

GOLDEN JUBILEE CELEBRATIONS

10th to 12th May 1961

PROGRAMME

Registered Office:
28 VICTORIA STREET, LONDON S.W.1
Telephone: ABBey 6672

Left, opposite and overleaf: **When GT3** was chosen to appear at the Golden Jubilee Exhibition of the Institute of Locomotive Engineers at Marylebone in May 1961 all seemed set fair with rising hopes that a production order would soon follow. The Duke of Edinburgh (seen here opposite left photo with John Hughes to his left) was an interested visitor as was William Stanier and the author, who was nearly ten. But this proved a high spot in the engine's life because cancellation and scrapping soon followed ending fifteen years of work, but not before the engine had briefly shown what she could do during trials in late 1961 (Overleaf bottom photo). (JH)

Colour Section • 79

Chapter 3
A SLOW BURNING AMBITION

In November 1947, having observed the way experimental turbine driven locomotives were being developed across the world, John Hughes, after pondering the whys and wherefores of the concept for many years, wrote:

> The Gas Turbine locomotive can be regarded as a development of immediate practical application, possessing a wider range of attractive characteristics than any available alternative for mainline haulage. The gas turbine mechanically connected to the driving wheels through reduction gearing is the only application in which the features of high power, and light weight of the prime mover are reflected in a locomotive of general simplicity and low first cost. In this form it is the natural successor of the steam locomotive and will, when satisfactorily developed, take the principal place in the world's railway transport.

For someone who had only just been recruited as a senior design engineer by English Electric, and so might be said to be feeling his way into this new post, this is a remarkably strong, ambitious statement to make. However, setting out one's stall in such a

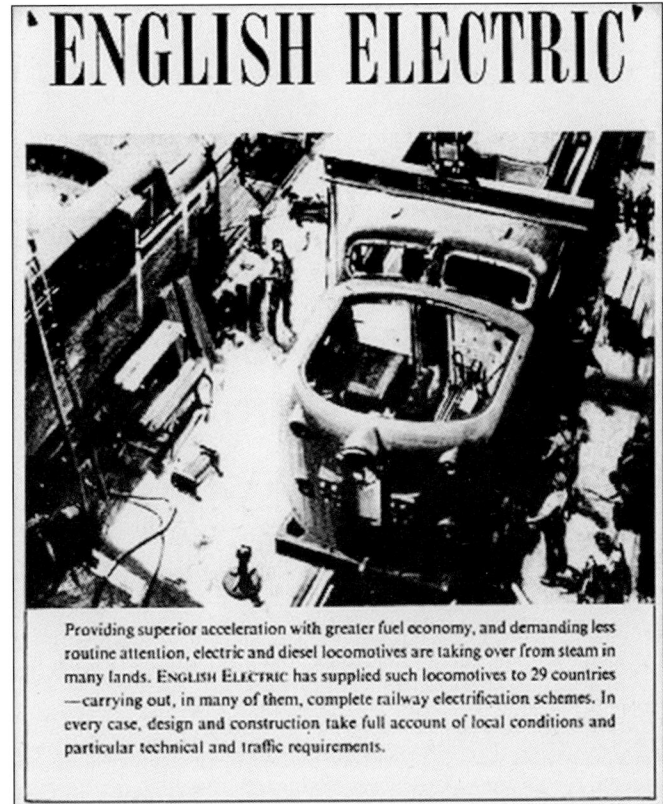 

Above left and above right: **When peace** finally returned in 1945 it brought with it a new challenge for English Electric because government contracts could not be guaranteed and there was stiff opposition for an ever-reducing supply of work. In this new world, railways, which had been starved of funds for years, seemed likely to create a burgeoning market for English Electric's goods. And so the company began to gather together many clever young designers, such as John Hughes, to drive these projects forward. The 1950s would see this speculative venture begin to pay significant dividends. (Author)

way runs the risk of making an individual seem overly confident and too sure of themselves. At worst, they may come across, as some did, as obsessive visionaries unprepared to pursue any end but the one they espouse themselves. In some cases, the end result may be a great success, but even when this happens, a more balanced, less certain approach, where many disparate but connected issues are demonstrably debated, is more likely to smooth the path to success. And should such a proposal fail, a clear audit trail in which all the issues are shown to have been addressed before a decision was made, is there to satisfy waverers and critics alike that the plan had been a sensible one worth pursuing. In 1947, no one, least of all his managers at English Electric, knew if Hughes was an overly enthusiastic visionary or simply someone with a clear view of the future. Either way, he seems to have gained the trust of his chairman, Sir George Nelson, very early in his career and this probably opened many doors that allowed him to gain approval for his developing ideas.

Once recruited by English Electric, Hughes quickly focussed his thoughts on turbines and played a leading role in the development of EM-27. When this project was nearing completion, he began the difficult task of convincing those with the power to decide these things that it could drive a locomotive. He did this by writing and then passing a number of papers to influential colleagues, so paving the way for what he hoped would gain approval to proceed. However, all this came at a time when many English Electric engineers were successfully arguing a strong commercial case for diesel engines and were making a compulsive case for the allocation of a large share of the company's limited resources to this work. In doing this they made great play of the fact that the LMS, sometimes in partnership with English Electric, had successfully initiated a diesel programme.

This had produced three diesel railcars in 1933, an active 0-6-0 diesel shunter development programme and, in 1938, an experimental, ultra-modern streamlined, three car articulated railcar set (Nos. 80000 to 80002). More was expected but the coming of war brought this work to a virtual halt. When meaningful work could begin again, post 1945, it did so under very restricted conditions – there

Sir George Horatio Nelson (above left) English Electric's Chairman from 1933 to 1962. Having trained as an engineer, he began work in Manchester with British Westinghouse in the reign of Edward VII. Here his undoubted skills were soon recognised and very quickly his career took off. In 1914, when only 27 years of age, he was promoted to be their Chief Electrical Superintendent and, in due course, was appointed manager of their Sheffield works. In 1930, he was chosen to be English Electric's managing director and became their chairman three years later. Over the next 29 years Nelson successfully built up the business making it a global force and, in the process, increasing employee numbers from 4,000 to 80,000, with turnover progressing at an even greater rate. He died in harness during July 1962 and was succeeded by his son Henry (above right). Nelson senior forged a close working relationship with John Hughes over fifteen years and was always, apparently, a strong supporter of GT3. Henry appears to have been less so and would be in charge when the project was cancelled. (Author)

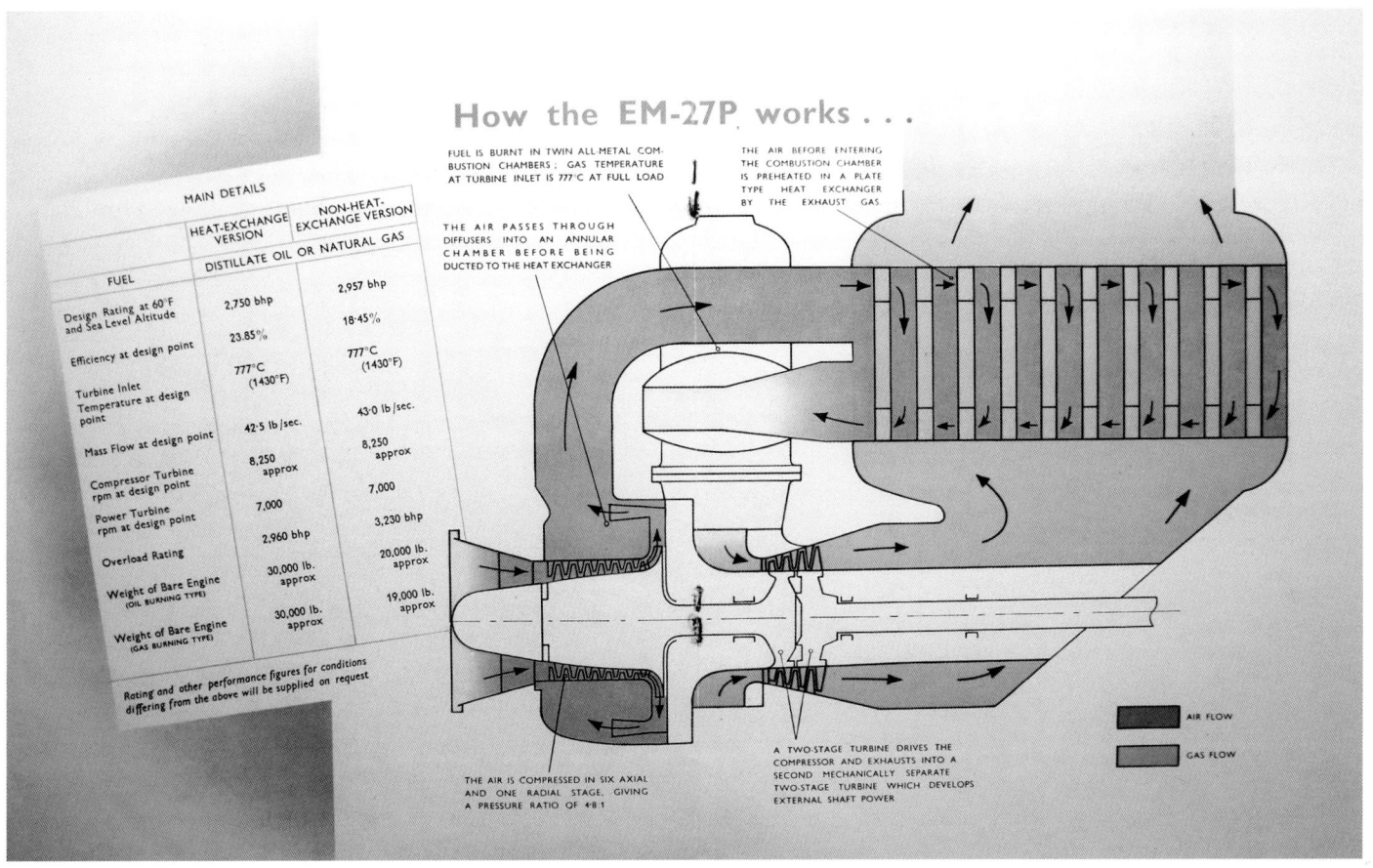

While English Electric considered future investments, and tried to reach a balanced view of emerging technologies to see which one might prove most profitable, they had already found a ready market in the burgeoning electricity generation business for turbines. The problem faced by those who wished to extend their use was building turbines much smaller than those in power stations so that they could be fitted into such things as aircraft and locomotives. The company proved particularly adept at this and in the late 1940s early 1950s unveiled their EM-27 2700hp gas turbine, a project in which John Hughes was involved. Very quickly, English Electric's PR Department circulated details of this invention to prospective customers (shown here). This in turn set in motion an internal development programme, led by Hughes, to build a locomotive around the EM-27. (JH)

being little money to do more than this. Nevertheless, sufficient was found to allow the LMS's last Chief Mechanical Engineer, George Ivatt, and George Nelson to marshal their resources and build two new mainline diesels – Nos. 10000 and 10001. The first of these engines was successfully introduced to service on the eve of nationalisation, hoping that BR's managers would be sufficiently impressed to order more, but this proved not to be the case.

In 1951, after much debate, BR initiated a programme that would see some 999 new standard steams engine built. This was in addition to the legacy projects they had inherited from the Big Four companies which were allowed to run their course. With so much invested in these programmes there was little available to spare for diesel, electric or turbine projects. With the benefit of hindsight, it was a decision that looks remarkably narrow-minded and uninspiring. Even France and Germany, where war damage was more extreme and there was probably even less money to spend on their railway systems, showed greater ambition and pursued diesel and electric options more fully.

In all honesty, there were some in BR who would have thought the steam policy unsound, but for the moment they lacked the voice or political clout to move the debate on. So, they bided their time, questioned the wisdom of continuing with steam and slowly prepared a case for a true modernisation plan. As they did so, companies such as English Electric, Brush, North British and Sulzer continued to plan a new generation of diesel and electric locomotives. In this they were helped in part by a growing interest in their products from railway companies in Brazil, Australia, New Zealand, Egypt, Holland, Nigeria, South Africa and India. This allowed them to experiment with new ideas and build locomotives that could then be tested and assessed in service.

Before the war, the LMS began to invest in diesel technology. In 1933 they introduced three diesel railcars and in 1938 an experimental, streamlined, three car articulated railcar set (Nos. 80000 to 80002), but the conflict brought this highly promising programme to an end, temporarily. With the end of war, under the direction of its Chief Mechanical Engineer, George Ivatt, dieselisation entered the vocabulary again. Work on two new mainline diesels, Nos. 10000 and 10001 (above), was brought to a conclusion on the eve of nationalisation. However, it was an opportunity wasted by BR's managers who ignored the needs of modernisation in favour of steam. Nevertheless, industry saw the writing on the wall and duly started investing in this or other emerging technologies. The problem that faced companies such as English Electric was in which direction they should go – diesel, electric or turbine – if they were to attract work from their biggest potential customer. In the event, dieselisation took precedence, but some such as English Electric kept turbine technology on the agenda in case its potential proved impossible to ignore. And, in due course, GT3 slowly backed into the limelight. (Author)

By 1954, the tide was turning in Britain and the end of steam was finally being predicted, though not always embraced. In December that year, the government published a long-awaited Modernisation Plan, with, at its core, two key ambitions – the introduction of substantial numbers of diesel locomotives for mainline service, at a cost of £1.24 billion and the phasing out of steam as these new engines began to arrive in service. And longer term it set a target of electrifying all the major routes, important secondary lines and many suburban systems. Here the plan called for the 1,100 new electric locomotives at a cost of £60 million, plus an additional £125 million to meet all other electrifications needs.

It was a very ambitious plan by any standards, and some felt that its scope too great so making it more likely to fail or, at least, face long delays. But at least it fired a starting pistol that allowed BR and a whole host of suppliers to begin work in earnest and provided the cash to do so. And at the forefront of this drive for change sat English Electric, eager to turn

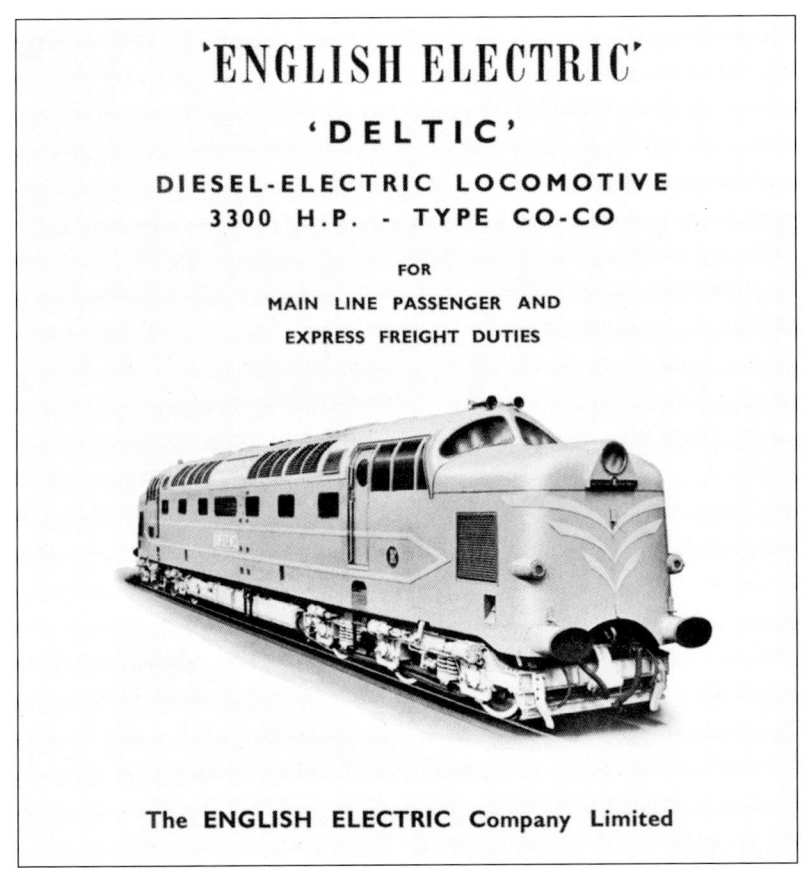

The BR Modernisation Plan published in December 1954 provided challenging targets and funds to pursue them. It meant the end of steam, but its emphasis on new technology gave many suppliers a chance to turn their experiments into working models. English Electric were quick to grasp this nettle and in 1955 produced their prototype Co Co configured DP 1 for BR to test. At the same time, they began working on four other diesels that would soon enter service – the Class 20 which first appeared in 1957, the Class 40 in 1958 and the Class 33 and the Class 37 in 1960. But the release of government funds also allowed English Electric to actively consider a gas turbine locomotive, so giving John Hughes the opportunity he had long been seeking to exploit. His was probably the 'poor cousin' to the diesel projects, and so well down the pecking order when it came staff resources and money, but it could be included in the modernisation plan nonetheless. (Author)

ideas they had been developing into working locomotives and, hopefully, substantial orders to help keep their immense workforce employed for the next decade or so. And with acceptance by BR that modernisation could not be long delayed, projects like English Electric's gas turbine engine, although of less importance than dieselisation and electrification, could take a step forward.

So it is hardly surprising that John Hughes, its chief advocate, was promoted in 1955 and appointed to head up its turbine locomotive development team. However, it is true to say that by this time he'd spent at least eight active years seriously considering such a locomotive and the problems to be overcome if it were to become a reality. These years of waiting hadn't been wasted, as his papers reveal. It was a time of thought, experimentation, collecting evidence, preparing a reasoned argument and enlisting support for the day when he could actively begin building his locomotive.

To achieve this, he began by offering this tantalising glimpse of might be gained from developing a gas turbine engine. In a paper produced in late 1947 he wrote:

Fundamentally, the locomotive suffers from two inherent limitations which are complementary. It has to haul its own weight, and that weight must be sufficient to produce

In the early 1950s, BR initiated a public relations exercise to help demonstrate how nationalisation was helping the railways modernise. This poster was one visible result of the this and made great play of the range of the new standard steam locomotives that were being built. Almost as an afterthought, five diesel, two gas turbine and two electric locomotives are included as though they were a major part of BR's plans. In reality they were all inherited from BR's constituent companies and were mostly experimental in nature. Only with the December 1954 published Modernisation Plan did the government finally initiate a programme based on more modern methods of traction. (RH)

the necessary adhesion for the driving wheels. In its simplest form, therefore, the ideal railway locomotive is one where the weight is approximately four times its tractive effort and all the weight is carried on the driving wheels.

So far every increase in tractive effort has tended to add to a locomotive's weight and its length; while at the same time the speed at which it is required to operate has risen too. Speeds bring in the question of riding qualities which must be such to allow a locomotive to use track of a given standard without undue wear and tear to itself or the track....The prevailing conditions in Great Britain impose a limit on space and weight of approximately 9ft x 9ft x 70ft and 120 tons in weight.

When assessing how a diesel engine meets these needs, by comparison to a gas turbine mechanical locomotive [Hughes believing that a gas turbine electric version, as built by Brown-Boveri for use in Switzerland, had significant disadvantages in weight and power] the turbine has the following main advantages:

Reduced weight for greater power.

Pure rotary motion and smooth torque.

Absence of water.. A ratio of standstill torque nearly three times as great as the full power torque of diesels, but this is only true where there is a separate power turbine.

He then went on to elucidate his case by comparing diesel and

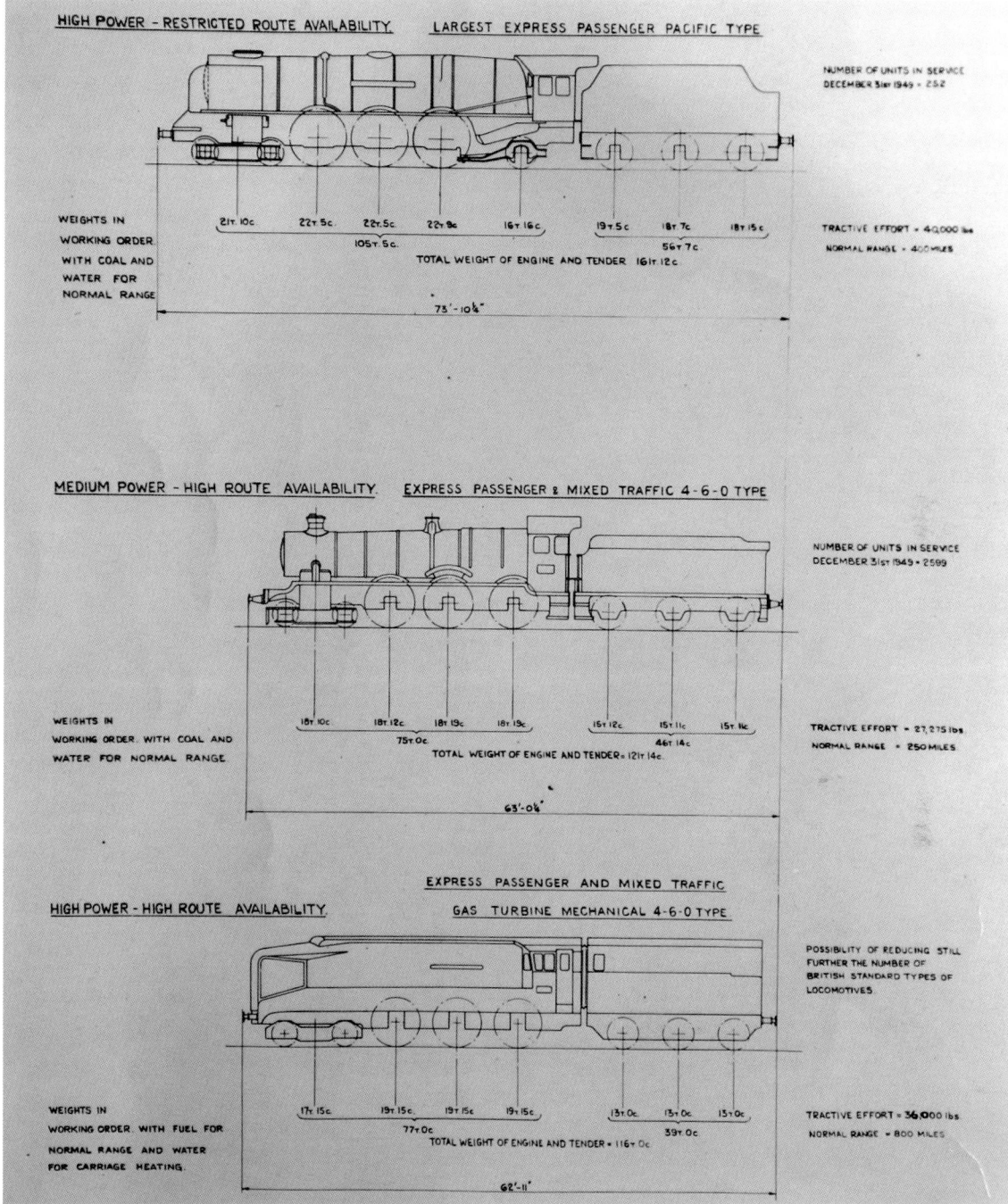

Shortly after being recruited by English Electric as a design engineer, John Hughes produced a paper outlining a proposal to build a prototype gas turbine engine. By then he had long been considering the idea and even produced some outline designs. The drawing above was attached to this 1947 paper with the aim of making a simple comparison with two of the successful steam locomotives then in operation. Even at this stage the eventual shape of GT3 can be discerned suggesting years of work on technical issues but also the engine's aesthetics. (JH)

turbine electric options with a turbine mechanical solution:

The gas turbine set which has been proposed is approximately one-third of the weight of a diesel of the similar power. The electrical transmission of a turbo-electric locomotive is heavy and represents 31 per cent of the total 120 ton weight which makes it unduly heavy for the power it gives at drawbar. Direct mechanical transmission will reduce the weight by 24 per cent and the length by 23 per cent.

Much of the power available to diesel electric locomotives will not be available at the drawbar because it will be absorbed in pulling the weight of the locomotive itself. The gas turbine mechanical will, by comparison give the most favourable figure for length and weight. It is estimated that a locomotive powered by an EM-27 2700hp unit will be 50ft long and will weigh 92 tons fully loaded. Its weight will therefore be 16 per cent of the total train weight, assuming a sixteen coach train of 500 tons tare.

Where the traction motors of the diesel electric and turbine electric are axle hung and nose suspended, the unsprung weight is high….. By comparison the direct mechanical transmission gives a comparatively low unsprung weight being free from dynamic augment. A number of drives have been developed for electric traction motors of much less than 2700hp in attempts to overcome the defects of the axle hung motor with mixed success.

The gas turbine has far fewer working parts than a diesel engine, and no rubbing parts other than shaft bearings, which are relatively free of wear. When the gas turbine mechanical is developed it will have greater reliability and will require less maintenance than the diesel… There is the further advantage that the gear transmission of the direct drive mechanical locomotive has 35 per cent fewer gears than are used in the axle drive of the electric system.

The weight and complications of diesel electric locomotives appear as a high capital cost which may be four times as great as that of a steam locomotive…The gas turbine electric is only at an advantage in the matter of cost over the diesel electric locomotive as long as the electrical transmission can be made to transmit the greater power of the turbine. Recent investigation into the gas turbine electric locomotive shows that within the British loading gauge and with all the present electrical practice, anything more than 2000hp can only be achieved with great difficulty. The gas turbine direct drive locomotive of 2700hp, with an overall length of 50ft and weight of 92 tons, which is still in the early stage of development, has a projected transmission of 20 per cent of the weight and 20 per cent of the cost of the electrical transmission it replaces. So the cost of the whole locomotive will be more favourable because the vehicle is lighter and smaller than a locomotive with electrical transmission."

Hughes concludes his paper with a detailed list of all the conditions his gas turbine mechanical locomotive must meet if it is to challenge the dominance of steam. More importantly he believes, and argues passionately, that it offers a better solution than either diesel electric or turbine electric locomotives when the time comes that steam is consigned to the history books.

Considering when this paper was written, in relation to the development of diesel and turbine traction, his conclusions, with little supporting evidence, are speculative to say the least and so are open to challenge. Nevertheless, if his intention was to grab the attention of those in power, he probably succeeded. With the promise of greater efficiency than other forms of motive power, lower production and running costs and the potential for greater reliability, English Electric's board would surely have taken note. Yet, in some ways, Hughes's initiative came too early, with the effects of nationalisation still to be assessed, so there was little point in developing a new type of locomotive when no one had any idea in which direction BR might go. In the event they chose steam which made private venture investments, at a time of recession, particularly risky. So the potential of Hughes' research was noted and put to one side awaiting better times. However, with George Nelson's backing, he was allowed to continue developing his theories, on a part time basis at least, for the time being.

While he waited for these better times to arrive, his primary focus remained on developing the EM-27 with thoughts he may have had about designing a locomotive remaining purely theoretical. However, he pursued the conjectural in two separate but complimentary ways, by separating the issues of form and function. Normally in design engineering, shape follows on from the creation of a machine and all its internal workings. In this case, form seems to have taken a leading role, helped by a friendship he had developed with Bill Allen, of Chapel Street, Belper, during the 1930s.

Little is known about Allen and what there is has mostly been gleaned from the letters he wrote to Hughes. It seems he was a talented amateur artist or a professional graphic designer, either way he clearly had some schooling in art, but also engineering and shared a passion for railway locomotives and modelling with his friends. He was born in 1893, the son of a railway Inspector, and saw service in the Great War as an officer with the Leicestershire Regiment. It seems likely that his and Hughes' friendship began when they both worked for Rolls Royce in their car division, post 1937. Here Bill's proficiency as an artist would have had a ready outlet in support of the company's many advertising campaigns and he would have worked closely with engineers to ensure his illustrations reflected their design aspirations. Confirmation of this is suggested by a letter Allen wrote in 1948 which alludes to work he regularly did for Mulliner Park-Ward of Willesden – one of Rolls Royce's partners – in this case responsible for coachbuilding their cars, as well as those built by Bentley and Alvis.

Park, Ward-bodied Phantom III Rolls.

Although their frequent letters make mention of Hughes' ambition to build a turbine powered locomotive at some stage, it isn't until 1947/48 that he clearly states his intention to do so. By this stage, his turbine development papers had been given wide circulation amongst English Electric's senior managers and George Nelson in particular. Encouraged by their response he then engaged Allen in designing the engine's exterior. A letter written by Allen on 13 May 1948 captures a flavour of the issues they discussed:

I am very much obliged to you for your letter, together with prints and photos that will enable future work to proceed on a more accurate basis.

From your previous letter I judged that speed was so vital that I am now able to enclose coloured drawings of 4-6-0 and the 2-6-6-2, whilst the 4-8-4 is almost complete and will follow during the holiday. These drawings may or may not meet

It seems that Bill Allen and John Hughes met when both worked for Rolls Royce's car division in the late 1930s on the Phantom III – Hughes helping create its engine and Allen its exterior design (above). Subsequently, they worked on several other models for Rolls Royce and Bentley, with coach work produced by Mulliner Park Ward in London. It was a partnership that proved key to the development of GT3. (Author)

requirements and I can easily do them again.

It is very kind of Mr Cheshire [believed to be head of English Electric's Gas Turbine Department] to consider a grant to me for this work and I sincerely hope that my efforts will be sufficiently useful for this possibility to become fact.

The three plates are mainly designed for use in 'selling' the projects and in any further work will be accompanied by scale drawings of a selection of descriptive matter. In view of this, will you kindly comment on the following items, firstly dealing with the enclosed illustrations:

1. Are they the right size?
2. Is the colour of the locos suitable?
3. The 4-8-4 is depicted handling a 'fitted freight' – is this acceptable?
4. The 2-6-6-2 is rather a queer animal and I have kept it austere in view of its humble role – do you agree?
5. Is the 2-6-6-2 equipped with two complete sets of intakes, filters etc? In my picture I have assumed not and that she is running with the intake trailing.
6. The 4-6-0 intakes were originally scoops – like jet aircraft intakes – but this would cause filters to be too far from the nose so I have just modified them in accordance with your lines on my pencil sketch.
7. The grill on the roof is thus at one end of the exhaust gas outlet trough and is similar to the grille on the model of the 4-8-4.
8. The tender is a bit of a problem on the 4-6-0 due to my aim to provide some rearward vision. I have kept the roof of constant width and section to line up with the cab and coaching stock,

A key part of John Hughes's work when designing and building GT3 focussed on wheel configuration and the engine's overall shape. In fact, deciding its form was a dominating theme long before detailed technical drawings, showing how the locomotive would function, were prepared by English Electric's draughtsmen. In correspondence dating back to 1937, when Hughes was employed by Rolls Royce, he had developed his thoughts on this issue with his friend called Bill Allen. Allen it seems was a graphic designer who shared an interest in locomotives with Hughes and turned their combined thoughts into a veritable production line of at least a hundred or more sketches, drawings and paintings (three of these are shown above). This work went on until GT3 appeared in its completed state. (JH/WA)

but incorporated the chamber in the upper tender side.
9. The all enclosed cab shown may have two serious disadvantages, viz: in bad weather closed up, the crew may miss a signal or become drowsy due to the heat – what do you think?
10. The panel work is a big problem. The only design advantage may be the fact that the panelling need not be as light as possible.

I have been wondering if the various radii – which can mean a great deal to the appearance of these 'new look locos', can be made from castings. Although mild steel panels would be required, I cannot at the moment decide what would be the cheapest and most suitable material for these castings.

The correspondence between them continued almost daily throughout May, with many design issues discussed at great length before being accepted or rejected (Appendix 1 covers many of these issues in greater detail). It is as if

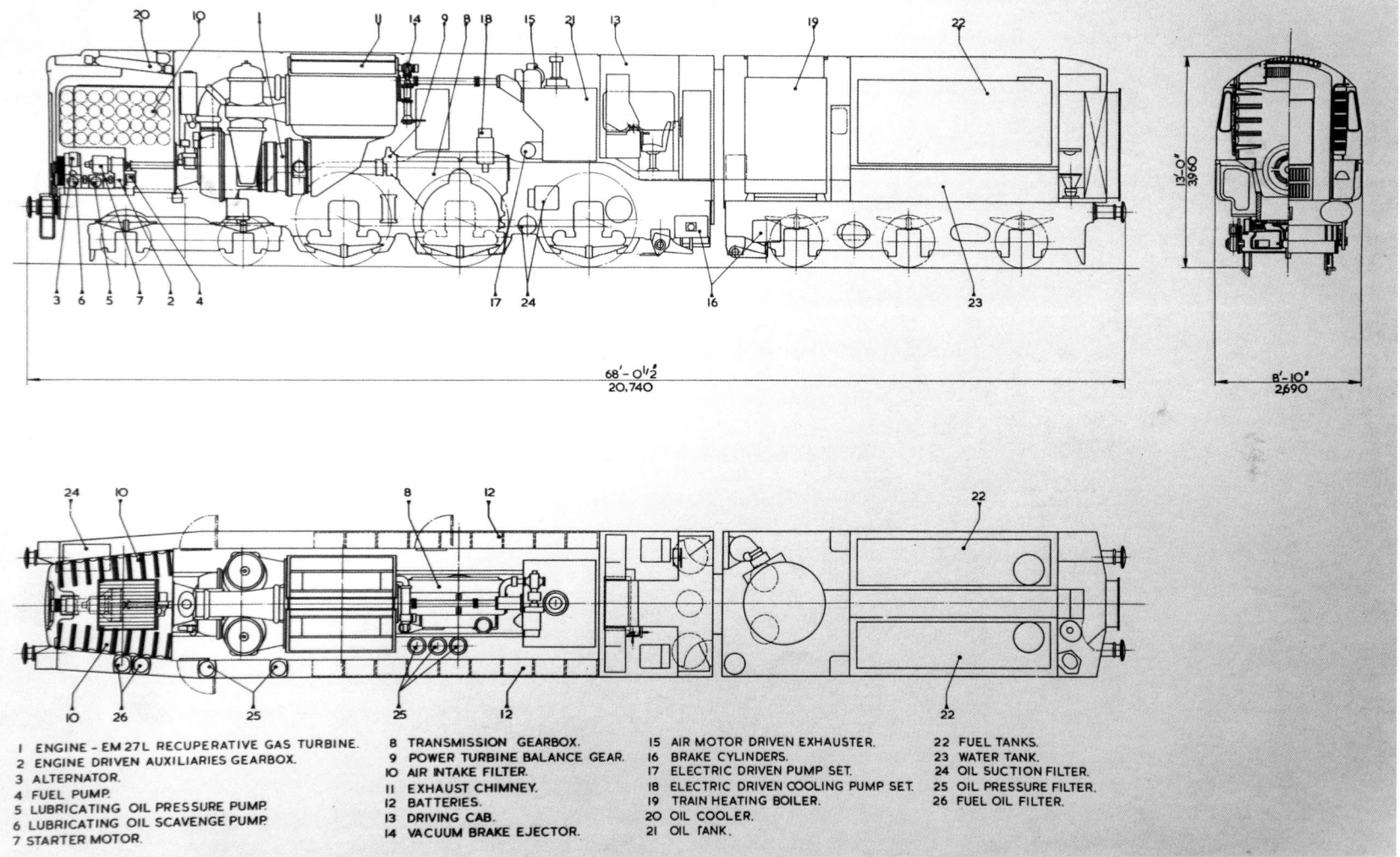

1 ENGINE – EM 27L RECUPERATIVE GAS TURBINE.
2 ENGINE DRIVEN AUXILIARIES GEARBOX.
3 ALTERNATOR.
4 FUEL PUMP.
5 LUBRICATING OIL PRESSURE PUMP.
6 LUBRICATING OIL SCAVENGE PUMP.
7 STARTER MOTOR.
8 TRANSMISSION GEARBOX.
9 POWER TURBINE BALANCE GEAR.
10 AIR INTAKE FILTER.
11 EXHAUST CHIMNEY.
12 BATTERIES.
13 DRIVING CAB.
14 VACUUM BRAKE EJECTOR.
15 AIR MOTOR DRIVEN EXHAUSTER.
16 BRAKE CYLINDERS.
17 ELECTRIC DRIVEN PUMP SET.
18 ELECTRIC DRIVEN COOLING PUMP SET.
19 TRAIN HEATING BOILER.
20 OIL COOLER.
21 OIL TANK.
22 FUEL TANKS.
23 WATER TANK.
24 OIL SUCTION FILTER.
25 OIL PRESSURE FILTER.
26 FUEL OIL FILTER.

When the first draft of this diagram was issued in 1949 the overall shape of GT3 had finally been established, coalescing in a way that reflected the outline and purpose of its inner workings. Up to then Bill Allen and John Hughes had been able to allow their imaginations free rein to add details that were, perhaps, more to do with form than function. In this process GT3's front end had, in particular, come in various streamlined forms, adopting some of the principles used in pre-war years by Nigel Gresley most famously on his A4 Pacifics, along the way. The outline shown above would, by the time GT3 was built, change very little. (JH)

John Hughes, as well as being a good engineer, was also, it seems from the papers that have survived, a good salesman. From his earliest days with English Electric, he knew how to put a good case together and present it in an effective way, even in model form as shown on this occasion with the name *Lord of the Isles*; He also knew how to get the attention of his Chairman, Sir George Nelson, and involve him in the project, even to the extent of having his name emblazoned along the sides of another model of the proposed locomotive. Whether this blunt instrument to appeal to his vanity worked is unclear, but Nelson seems to have remained Hughes's ally until his death in 1962. (JH)

Hughes was working to a tight timescale, suggesting he'd been tasked to do the work by someone in authority who was eager to explore the commercial possibilities of turbine engines. And, of course, by then English Electric would have known that the GWR, and then its BR Western Region successor, were pursuing similar ideas with Brown Boveri and Metrovick. So there may have been an element of competition in their actions.

By the end of 1948, the correspondence between Allen and Hughes appears to have dried up, insofar as the turbine programme was concerned. This suggests that George Nelson, who seems to have been actively involved in Hughes's work since his arrival, had thought again and decided to wait until BR's development policy had become clearer and funds became available for locomotives other than those powered by steam.

Hughes and Allen would continue working together in the years that followed, slowly refining many features of 'their' locomotive's external features, but it seems to have become more hobby than a project likely to change the course of railway history. In time it would be revived when Hughes's efforts finally bore fruit and, in 1955, he was given permission to begin working in earnest. In the meantime, he pondered issues concerned with the engine's internal workings and drew up a series of rough plans of layout and structures.

One issue that aroused much debate within English Electric throughout GT3's design process centred Hughes's strong advocacy of mechanical transmission over an electrical version. In this he was running contrary to popular opinion at the time, so trying to understand his reasons for doing so are of great interest. Yet in his seminal paper to the Institution of Locomotive Engineers during December 1961 he presented little or nothing on the subject, focussing instead on general technical details and how the engine was tested. Luckily, he added a little more in notes he prepared for a serious of lectures he planned to give, but appear not to have taken place:

Following the war there was a time when high hopes were entertained for gas turbines in many fields due to its overwhelming success in the air. In the first instance it was seen by English Electric as a more powerful and lighter equivalent to the diesel engine. It was this issue that I pursued with some vigour when joining the company in 1947. Fortunately, other engineers in the newly formed Gas Turbine Division of the company, several of whom had been directly involved in the work of Sir Frank Whittle on jet engines, or aviation, were well aware of the advantages to be gained by using turbines.

So, when one of them suggested that the better course would be to use the free power turbine with mechanical transmission for traction purposes, the proposal fell on receptive ears. It was realised that this option had not been pursued elsewhere and that it made possible a big advance

in this type of locomotive. Not only in transmission efficiency, which the mechanical system gives, but because of its general simplicity, particularly in the control system.

However, there was no general acceptance of this proposition within the Company and the inevitable committee was set up to report on these conflicting views. In due course it reported in favour of mechanical transmission, and the British Transport Commission supported this conclusion and said that it was a very promising line of development. As a result, my department was tasked with preparing designs for an engine supported by mechanical transmission during 1948.

Although the work was proceeding a conflict over its continuation and the nature of the transmission was not resolved and we constantly had to turn our attention from the task in hand to secure the project's future. This added significantly to the time taken to bring the project to fruition and meant it was subject to frequent review and the threat of cancellation. We were lucky that Sir George Nelson, our Chairman, was a strong and able supporter."

It was at about this time that a decision was made by English Electric to produce a turbine locomotive with a 4-6-0 wheel configuration. Others, such as the 4-8-4 and 2-6-6-2 versions, remained under consideration, but only in so far as they might be built at some unspecified time in the future when GT3 had proven itself in service. This, it was felt, might encourage BR to look more widely at developing a fleet of such engines for freight purposes.

Hughes later explained this decision, albeit briefly, by saying that:

In going for a 4-6-0 configured engine we had the potential of making the transition from steam easier for those who regretted it's passing – its layout being very similar. A turbine mechanical locomotive also provided greater power to weight ratio than the Western Region turbines or the new LMS mainline diesels. A 4-6-0 configuration provided greater adhesion and, at the same time, there was the issue of cost to consider. Building on

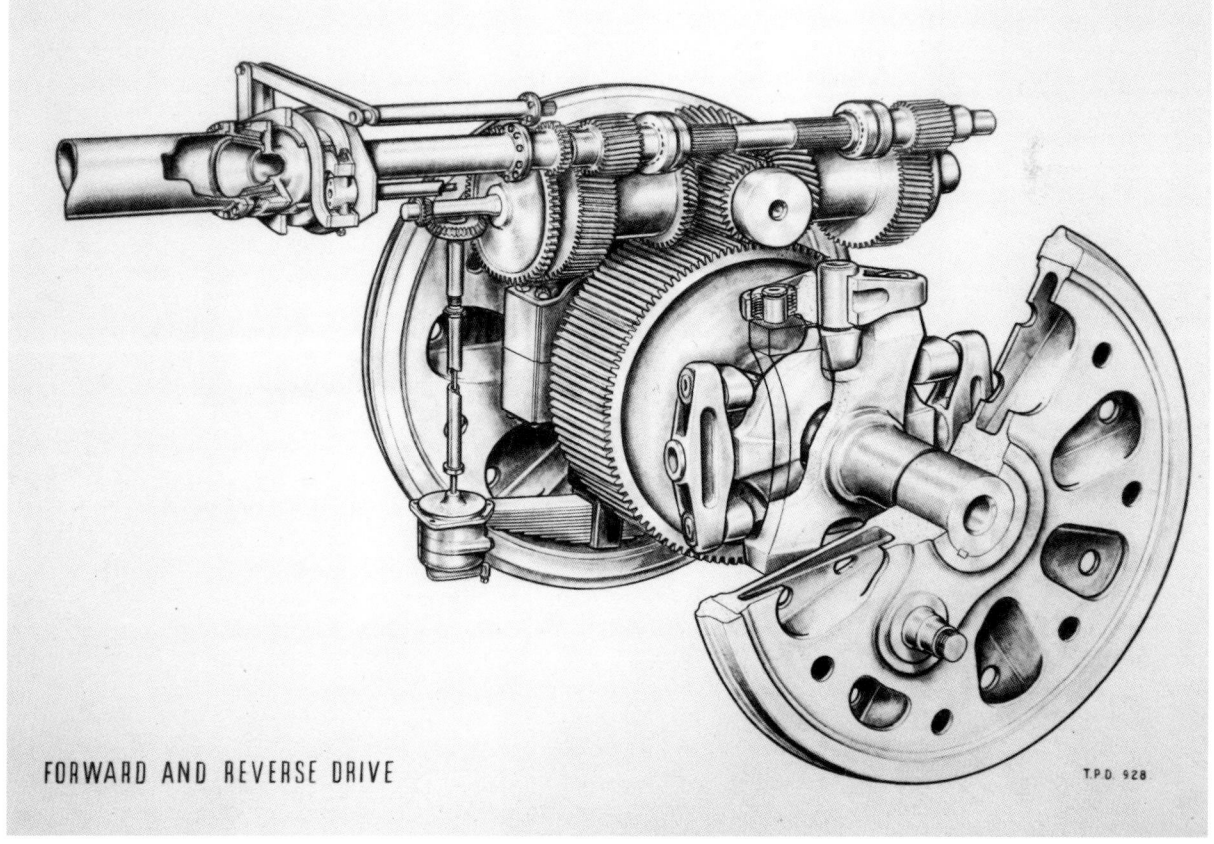

The mechanical drive and gearing of GT3 was a key feature of the design and according to John Hughes' 'owed much to that developed by the LMS and Metropolitan-Vickers of No. 6202 *Turbomotive*. By arrangement with the BTC we were able to study this locomotive.. Because the condition of the gears and transmission was of great interest to us, having run some 500,000 miles by then, we benefitted greatly by examining them at close quarters. The influence of No. 6202 on the design of GT3 was immense especially when it came to the simplicity of the solutions employed overall and in the control interlocks in particular and then there was the problem of misalignments in the drive it helped resolve.' This 1949 drawing captures the simplicity of GT3's drive, but also Hughes's desire to use Boxpok wheels similar to those applied by Oliver Bulleid to his Southern Region Pacifics, 0-6-0 Q1s and his Leader Class. English Electric didn't pursue this option when the time came to build GT3, preferring instead to use traditional spoked wheels. (JH)

existing, well tried frames and wheels would be significantly cheaper than an entirely new built engine. And of all the steam locomotives in service the 4-6-0 types were thought to be the most effective so in copying them, as far as possible, allowed us to build on a known and very successful entity. You only had to look at the GWR Kings and Castles and LMS's Black Fives to see the wisdom of following this path when building our experimental turbine locomotive.

Only time would tell whether this decision was a wise one or not, but even in 1948/49 the signs were not good. The LMS, with English Electrics assistance, had already adopted a Co Co wheel configuration for its two new mainline diesels and the two new classes of electric locomotives being built by the LNER had taken a Bo Bo form for its EM1s and Co Co for its EM2s. And in all three cases these engines carried cabs at both ends which had many advantages over steam locomotives when it came to safety and ease of operation. By configuring GT3 as though it was a steam locomotive probably saddled English Electric with an obsolete model which simply looked out of date in a modernising world. When trying to sell a new concept it probably helps if you are offering something that is not only technically sound and cost effective, but also looks contemporary. GT3 might succeed on the first count, but on the second it would probably fail – its quirkiness condemning it to be an also ran.

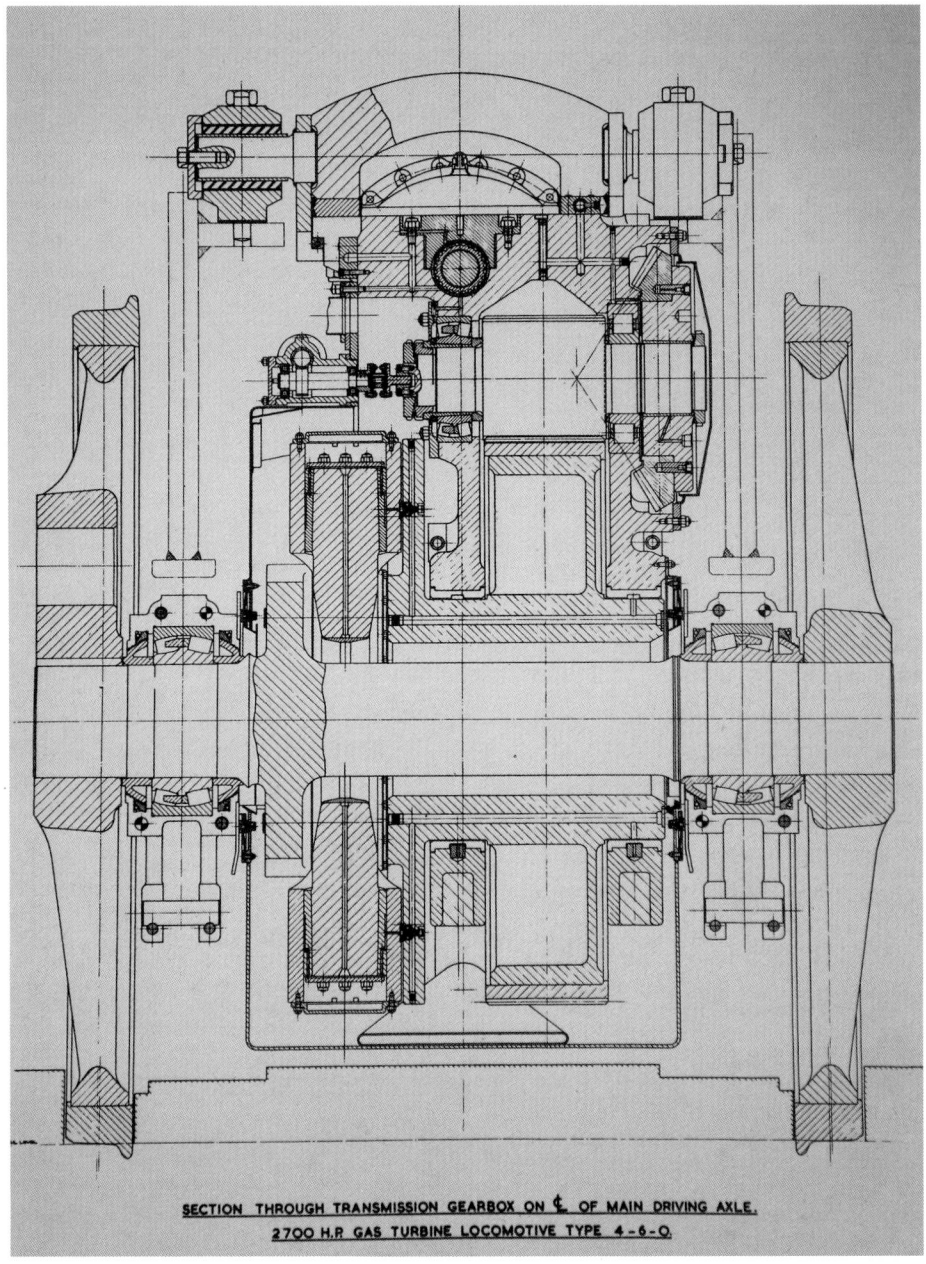

When considering all of the 1949 drawings prepared by English Electric, I was struck by the depth to which the company were prepared to go in producing designs for a locomotive that hadn't been approved and without a customer specification. I can only assume that Hughes and his fellow engineers in the Gas Turbine Department had sufficiently impressed those in power that GT3 was bound to attract interest when all the benefits in cost and efficiency were known. Sadly, this wasn't to be, and the project was allowed to languish awaiting the arrival of better times. Nevertheless, Hughes refused to allow his dream to die so easily and continued to pursue it as relentlessly as before. In these three drawings the transmission gearbox and the driving axle (above), the gear train (opposite) and the fuel system (overleaf) are given his treatment. (JH)

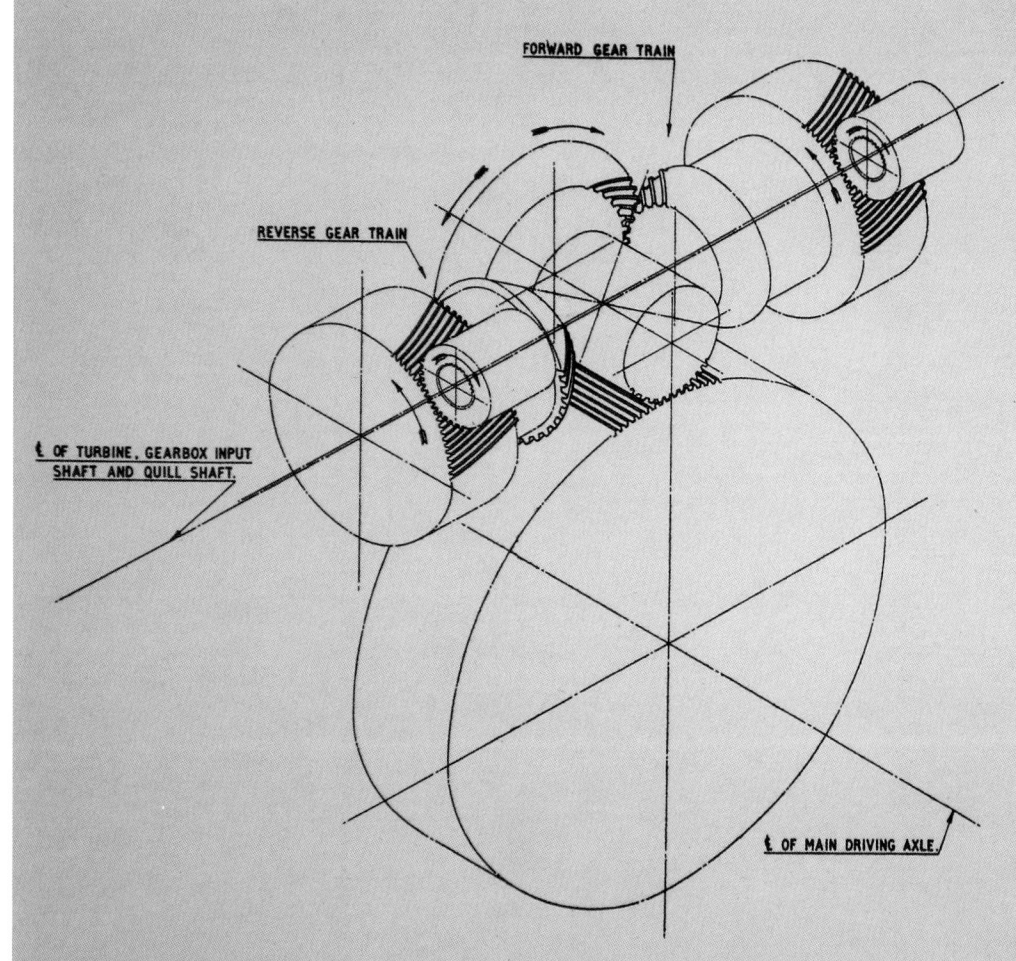

DIAGRAM OF GEAR TRAINS IN TRANSMISSION GEARBOX OF GAS TURBINE LOCOMOTIVE, TYPE 4-6-0.

Nevertheless, the decision had been made and GT3's eccentric form wouldn't undergo any significant alterations before being built. In 1948/49 when the project lost most of its momentum this probably didn't matter, but post 1955, when English Electric created its Gas Turbine Locomotive Project Group led by Hughes, it might have been wiser to change the overall design and adopt ideas that most engineers were already practicing, including English Electric. They didn't, which suggests that the company may not have attached any particular importance or priority to this issue. As a result, the engine's development seems to have attracted a fairly small budget, certainly far less than English Electric's gradually expanding diesel-electric programme. And yet it was allowed to continue, even when evidence that the Western Region's two gas turbine electric locomotives were struggling to make an impact, began to appear. Was this simply a result of John Hughes's compelling advocacy

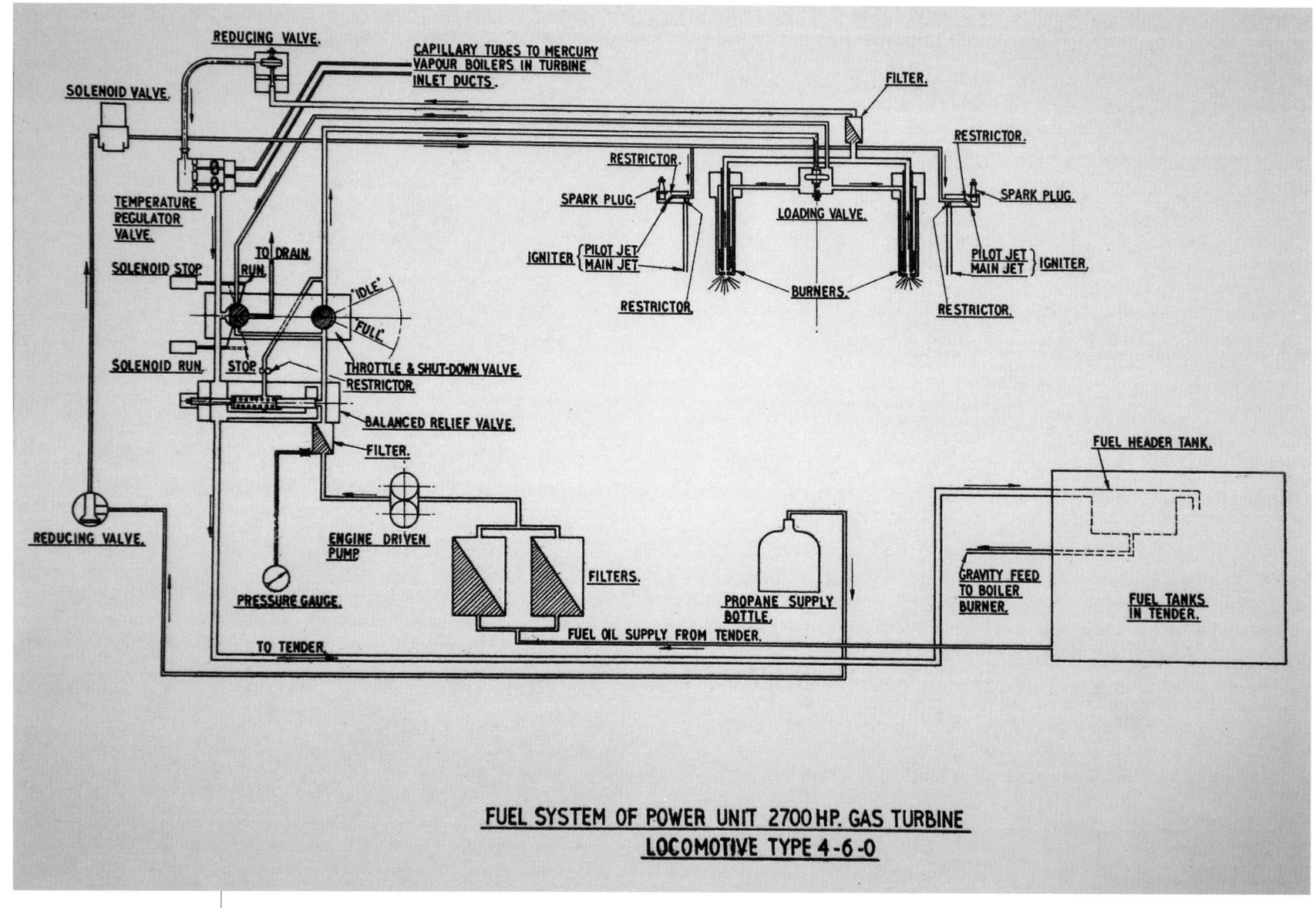

FUEL SYSTEM OF POWER UNIT 2700 H.P. GAS TURBINE LOCOMOTIVE TYPE 4-6-0

or was there a genuine belief that the project might eventually bear fruit? The truth would seem to be that English Electric, under George Nelson, were a company not afraid to speculate on new sciences and wager that the end result might prove worthwhile. Certainly in the past this had paid dividends, with successes outnumbering failures, and in GT3 a minimal outlay might do the same, so the gamble was at best a modest one.

So, the Gas Turbine Department, though thwarted in their first attempt to build GT3, were allowed to continue with their planning and hope that events ran in their favour. For John Hughes this meant other duties taking precedence and the new engine becoming a part-time occupation. But even in this mode he had much to observe and consider in the years up to 1955 when approval to proceed with construction would finally be given. When this happened, he produced a report for George Nelson and the Board that highlighted what he had been doing in the intervening period and what needed to be done if they were to succeed in this great venture.

His opening words were blunt and to the point:

> The present design of gas turbine locomotive (GT3) fulfils the listed requirements better than the Deltic, except in the cost of fuel. This objection will be largely overcome when the present engineering, to take advantage of the progress in burning heavy oils in gas turbines, comes to fruition. Fuel costing less than that for the diesel engine will be achieved on a future modification of the engine incorporating an all axial compressor. Construction of the test compressor from which this design will be evolved is now in hand.
>
> The development of the gas turbine locomotive to a total adhesion bogie type with mechanical with mechanical transmission, such as British Railway's desire, as contained in the recently published Modernisation Plan, can be foreseen along clearly defined lines.
>
> With success will come the demand for a range of powers, transmissions and applications that will eventually lead to the eventual disappearance of the diesel engine from a wide field of rail traction. In fact, the Deltic engine is seen as the zenith of achievement and complexity in diesel design, occupying a position similar to that of the great multi-cylinder aero engines of ten years ago, but destined to have a less dramatically sudden disappearance, in the more slowly changing world of the railways."

To say that Hughes was afflicted with missionary zeal in this matter may not be an understatement. But in the face of such determination and, with one eye to the untapped potential of gas turbines and its commercial possibilities, English Electric's directors allowed work to continue and, post 1955, to flourish.

During the years that GT3 remained in a sort of technical wilderness, John Hughes realised that the approach he must adopt if success was to be achieved must change. He wrote, in notes he later adapted for use in his 1961 presentation to the Institution of Locomotive Engineers, that:

> It was realised that it would be necessary to get a marked reduction in complication, and to achieve main and auxiliary equipment which would require less servicing and maintenance than existing types in order that the locomotive should be successful. This policy represented a big engineering effort since the design of the engine, transmission and auxiliaries would all to a certain extent break new ground. Because of this, it was important to avoid all other innovations in the locomotive which did not contribute directly to the end view and early in the design stage the type of locomotive chosen for the prototype was decided to drop the self-contained double-ended version and solely focus on one with separate tender in order to achieve our set purpose.
>
> The testing of a new type of locomotive with untried assemblies on a railway would be both lengthy and costly and could hardly be done without inconvenience to other traffic. So, a policy of design and test of components and assemblies was followed. And in 1955 it was agreed, with the British Transport Commission, that this might be better achieved if this work could be assisted by the staff and facilities at the Rugby Testing Station and the use of the rolling road there [which being set up for steam locomotives would suit GT3's 4-6-0 configuration very well]. With so much to do and with limited resources at my disposal it would take three years for the project to be sufficiently advanced for this to happen.
>
> It was in 1955 that I was also made aware of correspondence between Sir George Nelson and Robin Riddles [in effect BR's first Chief Mechanical Engineer] during December 1947 in which some of BR's objectives for such a project had been set out. Up to this time I was, to a certain extent, unsighted on this issue. But better late than never and Riddles' guidance allowed me to focus this final phase of the project on a detailed specification produced by our potential customer.

By the time John Hughes was finally given permission, in 1954/55, to begin building GT3, dieselisation and electrification programmes had become well-established. Though the number of each in service was comparatively small, by comparison to BR's steam locomotive fleet, they had begun to set the pattern for future designs – primarily Co Co or Bo Bo wheel configurations and cabs at either end. (Left – far right of the photo) English Electric's experimental Co Co DP1 Deltic, under construction at the company's Preston Works, surrounded by diesels mostly being built for export. The Class 55s descended from DP1 and began replacing the A4 and other Pacifics on the East Coast Mainline from 1961. (Above right) The LNER's prototype Bo Bo electric locomotive No. 6701 (EM1) that appeared in November 1940, one of a large number for the Manchester-Sheffield-Wath line. The coming of war meant the project was postponed, but in 1950 production resumed. The design of both DP1 and EM 1, and, for that matter the two Western Region gas turbine engines, identified a clear way ahead in locomotive design which Hughes seems to have ignored, sticking with his 4-6-0 base and cab located in a traditional steam locomotive position. (JH/RH)

The letter written by Riddles, dated the 18th, began with the words:

> I would say that the direct mechanical drive turbine locomotive is a very promising line of development, which is being worked on by various people at the present time. In this country it is known that the North British Locomotive Co and Messrs Power Jets Ltd are investigating the problem.

He then listed the key issues as far as BR was concerned:

Amongst the hurdles to be overcome are:

1. The gas turbine must be flexible and give reasonable efficiency at low speeds and low power. It should be capable of operating at any point in its power range continuously. If the gas turbine locomotive cannot tackle mixed traffic working it loses much of its appeal.
2. Life of the turbine and compressor blading must be thought of in terms of at least two years between opening up for inspection and overhaul – more if possible.
3. Life of the combustion chamber lining to be at least one month, and lining to be accessible for replacement in

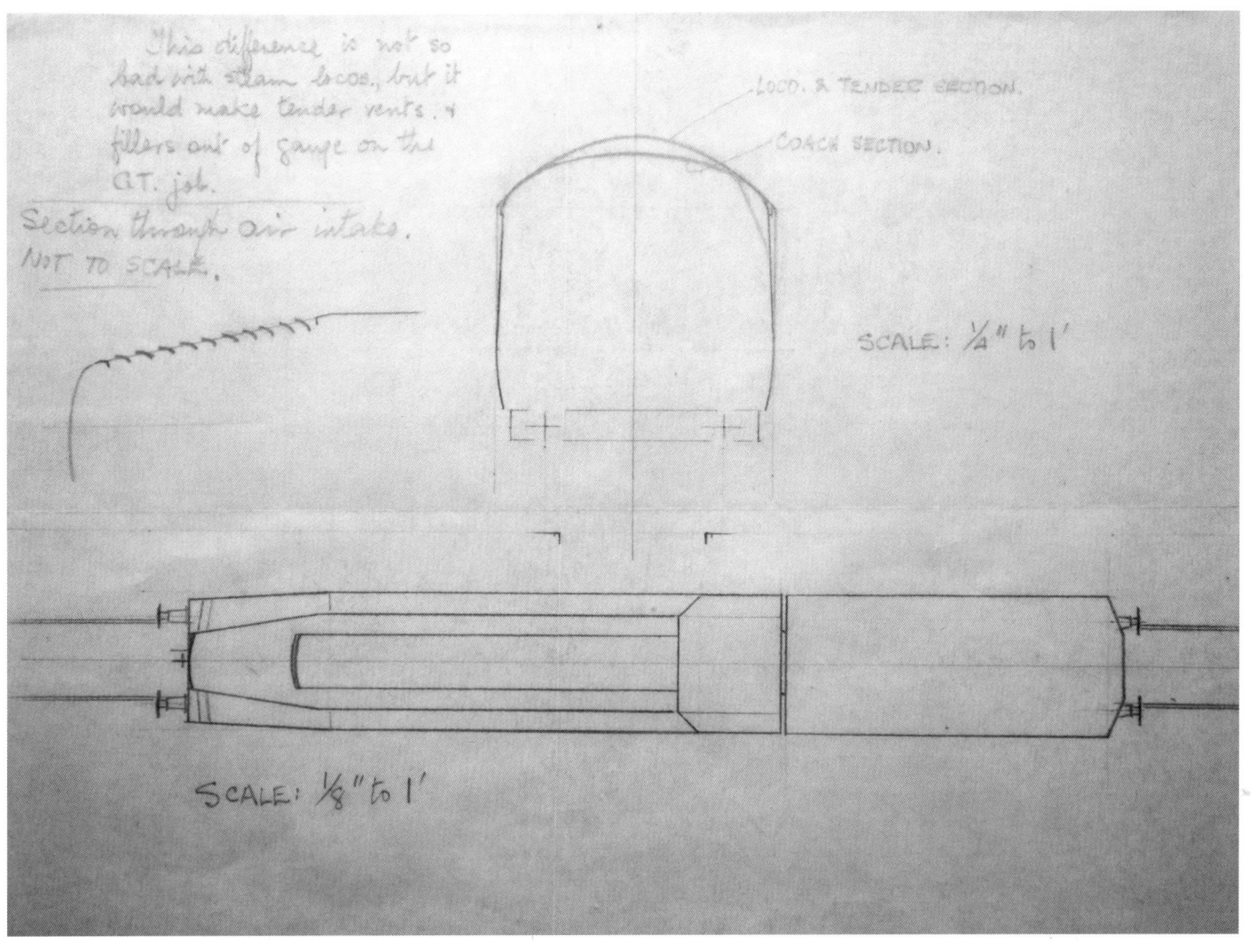

Even as late as 1957 Bill Allen and John Hughes were still debating the final shape GT3 might take. At this stage, as this drawing reveals, Allen preferred a more streamlined front end, but this would eventually give way to a more wedged shaped nose so that the air filters, oil coolers and other machinery might be adequately accommodated in the limited space available. (JH/WA)

the engine shed without cranes or special lifting tackle.
4. The dimensions of the turbine unit to be such that a through gangway can be provided from one end of the locomotive to the other.
5. Any links and pins in the final drive to be readily accessible for inspection and renewal without much dismantling.

Although these points seem to have been fairly obvious, contact with other gas turbine interests had shown that they are not always fully appreciated."

As Hughes prepared to finalise plans for the construction of GT3 he finally had some guidance from BR to consider and this, when combined with other already established design criteria, allowed him to press forward with his work. As his words make clear, he decided to do this on a research and development basis where both elements would be merged into a single, concurrent process. This is a common approach in complex, cutting edge projects where every component, let alone the complete machine, may still be at a theoretical stage of development. Having been involved in this type of work for many years it was a process well understood and practised by English Electric's scientists and engineers. This was especially so

It seems that GT3, although being approved in 1955, didn't attract the same resources as other projects, such as the DP1 Deltic diesel. With the extent of development necessary, and this lack of priority, the GT3 programme was a painfully long drawn-out process. John Hughes and his team were forced, by necessity, to proceed slowly, taking great care over the way individual components were designed and tested often without reference to other items of machinery to be installed in the locomotive. This meant that by the time it arrived at the Rugby Testing Centre, little was known about the way the engine might actually work in a completed state. As the project progressed, a detailed photographic record of each component's development was made, as these six photos reveal. (Opposite above left and above right) Drive components awaiting connection to the axle and the engine driven auxiliaries. (Opposite below left and below right) The centre axle components assembled with transmission gearbox and other equipment then added. (Top and above) A series of experimental air filters ready to be tested and the design finalised. The fuel drive components ready for assembly and testing. (JH)

on such things as the development of aircraft, avionics, missiles or other high-tech devices, which need many intricate and sophisticated systems to work effectively in a specific sequence if they are to succeed. But GT3, or for that matter, any other locomotive being developed for BR at the time, was unlikely to be considered in this category, so his approach may, to some, seem overly elaborate.

This suggests that by aligning the gas turbine project to truly high tech work that Hughes may have been overegging his case when seeking approval for the project to continue. But in this he was only doing what many others have done before, making a project seem more advanced and progressive than it actually is. If so, it worked up to a point, but in doing so he saddled himself with a burden of expectation that might be hard to live up to if and when the limited scope of the project was finally appreciated. Either way, it was destined to be a slow process and it wouldn't be until late 1958 that English Electric were able to get their locomotive to Rugby for testing and so benefit from the undoubted skills of the staff there. Sadly, by this time some thought it was too late for the project to truly impress BR or any other prospective customer.

In the meantime, Hughes and his team took the bare bones of their early development work and began to add much needed flesh to these bones. He also began the process of estimating some of the costs involved in the project. In doing so, his team produced an interesting comparison with

The first sign that GT3 was finally coming alive came with the cutting then assembly of the locomotive's steel frames. Being based on such a well-established 4-6-0 steam engine chassis carried with it few, if any, problems, so it is little wonder that John Hughes chose this solution from all the other wheel configurations he considered. It was a common-sense, pragmatic approach but one that probably rebounded on him later when his engine looked old-fashioned and hidebound in a rapidly modernising world. (JH)

steam locomotive construction, but not diesels, presumably because insufficient data existed, in 1954, when this report was written, to make this exercise worthwhile. His conclusions make interesting reading:

With a service based on the Western Region's Paddington to Plymouth route a comparison of costs has been prepared to include, a. Capital Cost, b. Utilisation, c. Fuel Costs and d, Maintenance Costs.

a. Capital Cost
Steam locomotive - £30,000.
Gas Turbine Locomotive - £65,700
The cost given for the steam locomotive is based on the known commercial cost of a Class 7 4-6-2 locomotive [thought to be an A4, a Princess Coronation or a new BR standard class Pacific – the paper doesn't make clear which it is]. The gas turbine locomotive is based on the projected costs of manufacturing all its component parts plus assembly of the engine – the complete turbine, gearbox and final drive, filters controls train hearing system etc.

b. Utilisation
The figures of hours per week and miles run, the mileage per year and the mileage between the major overhauls for the steam locomotive refer to the Western Region's Castle Class. The hours per week and mileage run for the gas turbine locomotive are those of that region's Brown Boveri engine. The mileage per year and periods between major overhauls are estimates by the Gas Turbine Department of English Electric based on what the design should achieve.

Steam Locomotive/Gas Turbine Locomotive
60 hours per week/77 hours per week.
1200 miles per week/2200 miles per week.
38 weeks – 45k per year/50 weeks – 110k per year.
90k miles per major overhaul/220k miles per major overhaul.

c. Fuel Costs
From diesel and gas turbine operations in service it has become apparent that the load factor is invariably much lower than the calculated all-out figure. For this reason, results from an actual service run from Paddington to Plymouth, with a dynamometer car,

As research and development slowly produced solutions, Hughes and his team began the business of assembling the component parts and trying to make all items of machinery work as one. This became increasingly more important the closer they came to sending the locomotive, in an incomplete state (as above), to the Rugby Testing Station for detailed evaluation. However, it was a process that became mired in difficulties. Hughes had originally hoped that the project would have reached this stage during 1956, at the latest, but the problems that had to be overcome ensured the trip to Rugby didn't take place until late 1958.

have been obtained from BR and will be used for the basis of these comparisons. It is assumed that GT3 will run on the same fuel as the Brown Boveri locomotive. If so, this will entail accelerating the research programme to make this possible. Up to now our development work has focussed on the use of distillates, operation on more easily and cheaply produced heavy fuel is likely to become the norm and until this happens, and we have evidence to show actual usage, no estimate of comparable fuel costs can be attempted (though heavy oil is likely to be far less than distillates). In due course, when we have these figures to hand, this will be added to the calculation.

d. Maintenance Costs
Maintenance costs for a comparable steam locomotive were obtained from the Western Region. These were divided into main works repairs and running shed

repairs on a mileage basis resulting in the following figures:

Per mile for main workshop repairs (steam) - 6d.
Per mile for running shed repairs – 4.2d.
Total maintenance costs per mile – 10.2d.

One third of the cost of main works on steam locomotives, and an even greater proportion of the running shed cost, is for boiler maintenance which is unnecessary in the case of a gas turbine. The total cost of gas turbine maintenance has, therefore, been estimated at 6.58d per mile, made up of main works repairs at 4.58d per mile and running shed repairs at 2d per mile (a breakdown of this estimate has been sent separately.

After some more analysis the author ends up with one recommendation for the English Electric's Board to consider, namely that, '£26k be allocated now to allow the development of the mechanical drive to be completed and tested and that increased effort be applied to the solution of the problems associated with heavy oil burning'. And to this he added a very telling note, 'if our work is successful we shall then be in a position to sell a prototype locomotive. There is evidence from interest shown by BR that a successful demonstration of the turbine and transmission would enable us to obtain a sale or return order for a prototype locomotive'.

However, Hughes was unable to disguise the fact that the figures he had presented, minus fuel costs, were marginally in favour of steam, which was hardly the ringing endorsement such a paper should have been championing. But in doing this, he missed one obvious detail – the capital costs he quoted for a steam locomotive were based on twenty, thirty or more being built, not just a single engine. By comparison, the gas turbine figure included all the research and development costs of building one experimental locomotive, which is a notoriously expensive business. Even the most junior of designers would normally have factored this into the equation and projected R and D costs being spread over a number of engines not just one, so achieving an important economy of scale.

Despite this, the funds requested were eventually and grudgingly forthcoming – not all were in support of the project – and the next stage of development could finally get underway. The first, and most important, step to be taken was to see if the EM-27 turbine could actually be adapted for use in a locomotive and survive all the stresses and strains movement would place on what was essentially a machine designed for static use. As things turned out, the lion's share of the £26,000 development costs, so recently bid for and allocated, would be swallowed up by this one issue. With a project appearing to be surviving on a shoestring budget, by comparison to other better financed schemes such as the DP1 programme, this placed a severe strain on Hughes's team. And as he himself later reported:

The work was proceeding but the conflict over whether or not it should be in English Electric's programme was not resolved. So it was not only technical difficulties to be overcome. Over the next few years, the engineers had to turn their attention, on many occasions, from the task in hand to secure its continuance.

For practical reasons, we decided at this stage that all work on producing the turbine, transmission and auxiliary components would be centralised at Whetstone, where I and my team were based. The remaining tasks – the production of the frames, wheels, superstructure etc would be undertaken at the Vulcan Foundry Works, where final assembly would later take place.

Cautious penny pinching and careful oversight of work are often a key to survival for any business, but for a research project, where the benefits may be guessed but not accurately calculated, it can stifle creativity and with it a desire to experiment. To save time and money, known solutions are often applied when original thinking might have paid bigger dividends in the long run. The fact that English Electric did not finance GT3 adequately suggests that some senior managers thought it had only a small chance of success, so the speculate to accumulate principle was underplayed in this instance. So, one wonders, if English Electric had been prepared to finance GT3 as liberally as other projects, whether it might have been

Anyone seeing GT3 for the first time was struck by its unusual looks, particularly its most striking front end. This gave it something of a 'space age' look with all those air filters clustered together on either side. Accortding to John hughes, 'this arrangement proved particularly effective because GT3 had huge lungs and the filters had to be so placed to make sure the turbine got the clean air it needed. This was particularly so in the 1950s and '60s with so much gritty pollution caused by all the residue of burnt coal hanging in the air. One forgets how dirty it was particularly on the railways'. (JH)

developed more effectively and completed more quickly. And, if so, whether this would have made the locomotive easier to sell to a main customer finally eager to move forward into the future. When GT3 was finally ready for service in the early 1960s, it was probably too late for BR to take seriously, by then very heavily committed to diesel and electric locomotion as a means of modernising its fleet.

Nevertheless, work proceeded with development of the turbine itself taking up much time and effort. The

106 • GT3 THE UNREALISED DREAM: THE STORY OF BRITAIN'S LAST GAS TURBINE LOCOMOTIVE

As GT3 was finally being assembled, the opportunity was taken to photograph the incomplete engine in the open air. John Hughes later wrote that, 'although there was much still to be done we took great pride in having got this far. When seeing the engine reach this stage of development we believed the worst was behind us and we hoped that once tested at Rugby and modified where necessary, it was but a short step to completion, with, hopefully, orders to follow.' The location of both these photos isn't recorded, but the date is August 1957. (JH)

central, most important issue seems to have been developing, as Hughes reported, 'a two shaft gas turbine, associated with a simple geared transmission, that would give the desired starting traction of 40,000lbs. Coupled to this it was decided that the turbine should, for an optimum running speed of 50-60mph in the range of 5400rpm, produce a standstill torque of 5600lbs.'

To develop the many theories on how this might be achieved, it was decided to build a scale model to test and assess the starting performance of the power turbine. To ensure optimum performance this model was fitted with blades selected from low speed wind tunnel tests with 'a variety of profiles'. As a result:

Fresh turbine calculations were done in which the optimum running speed was altered to 5000rpm. This being a compromise between the requirement for high torque at standstill and operation over a wide speed range, but sufficient confidence was felt for a full size turbine to be manufactured to this design

In August 1958, John Hughes also took the opportunity to take many close-up photographs of GT3 showing each part of the engine and how they fitted together. This picture shows the turbine in place to the rear of the air intake filters. (JH)

and tested with a dynamometer using air supplied by a steam turbine-driven compressor. The results were unexpected in that the swallowing capacity was reduced as its speed continued to rise. This was sufficiently great to increase the back pressure on the charging set and move the compressor operating line towards surge.

This made this turbine unsuitable for use in the engine because, for instance, should the locomotive coast down a gradient under reduced power, but at high road speed, the size of the power turbine would be small enough to cause the compressor to surge dangerously. In addition, the swallowing capacity was some 4 per cent too small and its peak efficiency was 6 per cent below the calculated value.

It was decided at this point to abandon the two-stage turbine and use the knowledge gained to design a three-stage version with the optimum speed of 5000rpm.

Having dealt with this central issue, which seems to have taken a year or two to resolve satisfactorily, there was still a good chance that the project might have been cancelled by the predictors of doom amongst English Electric's senior managers, so a great deal rode on this one element of the design. But, eventually, it passed muster and work on two other key aspects of the scheme – the combustion chambers and the heat exchanger – could move forward. There was much else to do apart from these three key elements, such as the gears, transmission, frames, engine controls, the generator and more, but Hughes believed that here:

> Well established, conventionally engineered solutions existed so making these elements of the design easier to resolve, with much resting on what we might discover when, and if, we progressed sufficiently to allow the locomotive to be tested at Rugby. Throughout 1956 and '57 this seemed increasingly unlikely as our work hit many unforeseen brick walls.

The combustion chamber issue was partially driven by BR's restrictions on the width and height of engines imposed by its loading gauge. So as not to fall foul of the limited space available, it was necessary to fit two combustion chambers not one. These were of the return flow type which were connected to the outlet ports of the heat exchanger by short ducts which 'embraced the barrels of the chambers eccentrically'. This was not an entirely satisfactory arrangement as Hughes later related:

> This solution resulted in the air which enters the mixing zone having a considerable swirl which even persists in the air passing to the top of the chambers so severely effecting combustion and, occasionally, causing it to fail. After much research, with various solutions being tried, it was found that the flame could be kept nicely within the rather short barrel by fitting guide vanes for the entry of the combustion air which imparts an equal and opposite direction of swirl to that of the air entering the mixing. Although when bench tested this solution seemed to be effective, we needed to see it working in fully stressed conditions on the locomotive, before judging whether it did what we hoped it would do. When eventually we reached Rugby testing revealed its flaws.

When it came to the heat exchanger the only way it could be fitted into the body of the locomotive, without compromising its performance, was to use:

> Plates for the heat transfer surfaces with channels pressed into them so that when welded to a neighbouring plate flow paths are formed for the air and exhaust gasses. Here a straight path was provided for the exhaust from the bottom to the top of the plates, whereas the air has to enter and leave from the side, so only a portion of the plate length gives a true contraflow conditions for the gases.

When under test the performance of the heat exchanger was disappointing as a thermal ratio of around 57 per cent was recorded instead of the design value of 65 per cent. In addition, cracks developed, mostly in the argon gas welds. The deterioration became so bad in the heat exchanger that it was eventually withdrawn and cut up for forensic evaluation.

The fault was found to relate to the welding method which was modified with a shield of argon applied above and below

the main weld and a change was made in the specification of the stainless steel used in manufacturing the plates. This information was used in the construction of a second unit, which was run without trouble during a 1,000 hour endurance test, then subsequent cycle trials. The only changes made to this version of the heat exchanger were external in nature to allow it to link into the other machinery in the engine.

It is interesting to note how simple it is to write and describe a snag that may have taken many months or years of experimentation and frustration to correct. And with each delay the cost rose and the time when a completed locomotive might finally appear drifted into the distant future. As the full extent of the project's difficulties became apparent, so the estimated cost of £60,000, so confidently predicted in 1954, climbed to £112,000 late the following year. It is hard to say how this massive increase was received, but those who had been critical of the project from the beginning were given fresh ammunition in their bid to bring it to an end and invest these resources elsewhere. Luckily, Hughes seems to have retained Sir George Nelson's support which must have effectively silenced any opposition, but only time would tell how long this might last.

After so much effort had been expended and with it some considerable sum of money, Hughes felt able to book GT3 into the Locomotive Testing Station in the early summer of 1957. Judging by the frequent correspondence dating back over more than two years, some at Rugby had begun to wonder whether this day would ever arrive. Nevertheless, when it did the engine was still far from ready and there was a scramble to finish a number of tasks in the few weeks before the locomotive was transported to Rugby in July. One of these was to:

Bench test the engine where we established that the output at the design maximum turbine inlet temperature was only 2100hp and the charging set

Very few photographs appear to have been taken of the team that developed GT3 and those that exist don't seem to have been annotated with names, dates or locations. (Above left) Some sort of official presentation to a member of staff which John Hughes (second right) attended. (Above right) One name that does occasionally crop up is that of George Hughes who was described as GT3's test engineer, but whether he was employed by English Electric or was a manager at Rugby is unclear. (JH)

In the weeks before GT3 was transported to Rugby for testing to begin there was a great rush to make the engine ready, with many essential tasks relating to the power turbine and the engine's controls to be completed. According to notes kept by John Hughes, this picture was taken at the Vulcan Foundry Works when Test Centre managers visited English Electric to discuss the programme and see for themselves the state of preparedness the engine was in. Hughes recorded that, 'after a close inspection and much discussion, Carling [Dennis Rock Carling, Rugby's Superintending Engineer] agreed to take on the task subject to certain items of work being completed to his entire satisfaction'. (JH)

In July 1957 GT3 was sufficiently complete to allow a very detailed evaluation to begin at Rugby's Locomotive Testing Station. Here the engine is photographed, according to John Hughes's notes, 'at the centre shortly before the first trial began with a mixture of my and the Superintending Engineer's staff posing in front of the locomotive'. (JH)

was operating some 450rpm lower than its maximum design speed. The overall turbine efficiency was found to be low, but due to lack of inter-turbine instrumentation at this time it was not possible to specify the exact cause of the lack of power other than that the turbines were mismatched. However, the power turbine was free from any unsatisfactory characteristics and in all other respects the engine ran satisfactorily and was deemed ready for a more detailed evaluation at Rugby to begin.

Hughes undoubtedly breathed a sigh of relief when finally reaching this stage of the project. To do so he had taken a long and sometimes troubled route, but now he could put into action the second part of his plan and let Rugby, in co-operation with his own staff, take the design process forward and help iron out any of the engine's shortcomings.

Usually, staff at the Testing Station simply evaluated a completed engine with well-established characteristics, but now they were being specifically asked to help in locomotive design, alongside English Electric's own engineers. It was, as Dennis Carling, the centre's Superintending Engineer, wrote, 'a unique project for us to be involved in and one we hoped would open up a new and possibly unique avenue of work for us to explore and so extend the life of the facility beyond the end of steam locomotion. Sadly, it was not to be and meaningful testing ended in 1959.' So, GT3's arrival that year had more than John Hughes' hopes riding on the outcome of this 'unique' exercise.

Chapter 4
OUT OF TIME

The Rugby Locomotive Testing Station as it appeared in the 1950s. This purpose-built facility probably came too late to be of real use and by 1957, with steam's demise on the horizon, it had an uncertain future. BR could have adapted it for testing diesel locomotives but chose not to do so, presumably because its managers preferred their locomotive suppliers to undertake this task before BR accepted their products into service. Whatever the reason, the Testing Station gradually fell into disuse and was demolished in 1984. It is probably true to say that the potential identified by its main advocate, Sir Nigel Gresley, remained largely unfulfilled. In some ways, GT3 became the facility's swan song, because little of any great significance apparently took place following the gas turbine locomotive's departure in January 1958, except for tests with a 5mt 4-6-0 and two 9F 2-10-0s. (RH)

After such a long, trying gestation, John Hughes must have hoped that test station staff might quickly validate all the work done and then offer advice on the way the project might be brought to a successful conclusion. There were many observations and suggestions made but the most telling comment involved how such a locomotive might be employed in the future should it attract customers:

The characteristics are such that the locomotive will show to best advantage if it is employed on duties involving sustained high power and mainly running between 40 and 70mph, such as heavy night trains with sleeping cars, fitted freight trains and such like. The basic concept is one of a machine for long distance working at high power.

Perhaps not the high praise he was hoping for, as he fought to keep the project alive, but it wasn't a damning indictment of his work either. But, as always, the devil is in the detail and in the days that followed Hughes and his team closely scrutinised the often nuanced data contained in the test report, carefully extracting the lessons learnt for further evaluation.

Later on, he would write a note to his managers which simply stated that 'The importance to the success of the project of the work undertaken at Rugby cannot be overestimated. It has provided a body of invaluable information and evidence that will inform and guide the next stage of the project.'

There then followed a more detailed analysis of Rugby's findings (which are recorded in full in Appendix 2), which do not appear to have been given wider circulation. Having seen the way some of his colleagues had used any perceived shortcomings as reason to cancel the project, this is, perhaps, not surprising. In the event, his paper seems to be a restrained and balanced account which would later be modified and form part of his 1961 paper to the Institution of Locomotive Engineers:

The test programme covered the running-in period of the gear drive, measurement of performance of the locomotive over a wide range of power and speed and determination of the energy losses in the transmission

John Hughes watches the driver running up the turbine in a static tests outside the Testing Station according to the notes left by Hughes, 'on a wet day in September. A temporary cab had been erected to provide some cover for the driver. This canvas shroud would later be replaced by a metal umbrella secured by six bolts to allow quick removal when entering the main hall to undergo rolling road tests'. (JH)

GT3 about to begin another run on Rugby's rolling road with, it is thought, John Hughes in a white coat on the footplate. During the many months of trials held there he was a constant presence at the Testing Station. (JH)

gearbox. The running in was done with progressive increases of load and speed, whilst a careful watch was kept on oil and bearing temperatures, with the gears being examined at regular intervals to see how they were standing up to the constant strain.

The running in period was achieved without significant difficulties, as far as the transmission was concerned, but there was a good deal of trouble arising from failures in the lubricating system. Here the tests showed that in certain circumstances the scavenge pump, which delivers the oil to the top of the tank via a cooler, to keep the oil level at the bottom of the sump below the level swept by the final drive, was ineffective. As soon as 'splashing' commenced above a certain speed the condition rapidly deteriorated until large

quantities of oil were forced into the upper part of the gearbox, with some escaping through the axle seals. A cure for this was eventually found by modifying the sump bottom which, after several other solutions had been tried and tested, brought the scavenge pump's suction pipe in from below instead of from above.

The lubricating system also caused us trouble when it was discovered that cold oil failed to de-aerate fully if the locomotive was opened up too quickly from the first start of the day. This caused a froth of oil to escape from the vent at the top of the tank. After much discussion with Testing Station engineers, and four failed modifications, the problem was finally cured by increasing the tank's volume above the normal oil level and fitting immersion heaters that would operate when the locomotive, after a spell of disuse, was started from cold.

After these initial running-in trials tests proceeded by gradually increasing speed, power and pull, first separately and then combined. The upper speed limit was set at 90mph, its designed maximum, with the lower limit set by either the capabilities of the test plant or the limit of adhesion. During the trials that followed there was a marked deterioration in performance caused by damage to turbine rotor blading as a result of contact with the stator casing. This limited maximum power output to about 80 per cent of that expected.

It was possible to make an estimate of energy losses in the transmission gearbox from the heat taken away by the lubricating oil, since the locomotive could not be run on the test plant at a constant speed long enough for thermal

As the trials ran their course and design flaws were revealed the engine was occasionally stripped down so that remedial or modification work could be undertaken. Rugby was not equipped as a workshop or a maintenance facility, so any work of this nature presented the team from English Electric with a number of practical problems, not least of all heavy lifts of major items of machinery. A road mobile crane had to be brought in to do this, with all work, as seen here, being completed in the open air. (JH)

equilibrium of the gearbox to be reached. At 60mph, for example, the transmission heat loss was 168hp and the cooler heat drop 163hp, which showed a satisfactory correlation. However, it was not possible to obtain a complete power loss curve, because at lower speed the heat circulating was low in relation to the thermal inertia of the system which meant that equilibrium could not be reached in a reasonable time.

It was clear from initial tests at Rugby that the power turbine was too small; so while provision was being made for inter-turbine instrumentation the engine was rebuilt with the first row of power turbine blades removed. This was a quick way of increasing the power turbine size without any loss of efficiency. Following more tests, it was discovered that the engine output had increased to some 2400hp, up from 2100hp, at the maximum turbine inlet temperature, although the charging set speed was some 250rpm above the design value. At the same time thermal efficiency had fallen from 21.7 to 20.2 per cent making it apparent that the power was set too low.

After much more work a means was found to determine the performance of individual turbines and a final set of tests undertaken. But before this happened the charging turbine blade tip clearance was adjusted and the tips feathered to reduce the risk of 'pick up'. In addition, the lacing wire to the power turbine stator blades was machined to give a better aerodynamic form and the exhaust chamber was increased in size to improve flow characteristics. These changes proved successful and brought the engine to its present state with an output of 2600hp at an overall efficiency of 22 per cent and a charging set speed of 8240rpm.

During the tests at Rugby trouble developed due to the distortion of some of the rings in the combustion chambers. This resulted in a large proportion of all the cooling air passing through to do so at one spot, so causing overheating which, in turn, led to a rapid destruction of the flame tube. A new design was prepared in which the perforated outer shell was replaced by a light cage arrangement that formed a fifteen-sided figure containing six individual louvred elements in each face made from alternate layers of flat and crimped strip welded together at their ends.

This work, plus some other minor modifications, has given very satisfactory results. After 500 hours of testing, with varying loads, there was no sign of overheating or any deterioration in the face of the louvred units, which showed every sign of having a very long time free of maintenance."

The list of problems uncovered during the months GT3 spent at Rugby, when the engine travelled some 5,000 miles on the rolling road, was not an extensive one. Nor were the shortcomings uncovered insurmountable, but to help ensure that all that could be revealed had been revealed a period of testing on the 'open road' was thought necessary. With the engine in an incomplete state this couldn't, for obvious safety reasons, be on the mainline so GT3 was trundled around 'tracks at the Testing Station and in the yard of Rugby Motive Power Depot and adjacent sidings'. To give these tests a semblance of realism, the loads pulled were made up of two steam locomotives with their brakes applied by varying amounts. The results must have made pleasant reading for Hughes:

Starting against a heavy load was tried by using two eight-coupled steam engines and tenders as the load and then starting on progressively more severe gradients. The most severe condition attempted was on a gradient of 1 in 44 combined with a curvature of 5 ½ chains on roughly laid track, when a start was made without slipping although the locomotive was, in its unfinished state, nearly four tons light in adhesive weight. The start was rather slow as there was an appreciable delay whilst the charging set speed was increased slowly until the torque on the power turbine rose to a high enough value smoothly, gradually increasing speed. The opinion expressed that no steam locomotive then using this line (Class 5MT and Classes 7F and 8F) would have been able to start the load at that point on the line is pertinent. The rails were dry but not sanded.

To simulate the sorts of loads GT3 was likely to pull in service, staff at the Testing Station adopted the simple expedient of using steam engines with brakes applied by varying amounts to create degrees of drag. In this case two ex-LMS 8F 2-8-0s were used and are here photographed during one of these tests on the siding alongside the Rugby to Peterborough line being pulled by GT3 up 'a sharp gradient with their applied brakes making an almighty racket', according to Hughes. (JH)

During these 'mainline' trials, the locomotive behaved fairly well, considering its incomplete state, and the only problem of note that arose concerned the reversing gear which failed to operate at times. The engineers, once the parts had been stripped down, discovered that, 'there was too much friction in the large diameter bearing carrying the brake backplate for sufficient rotation when necessary to allow engagement'. After some 'head scratching' it was discovered that by supporting the brake on fabric bearings then resting them on the gearbox input shaft, either side of the brake discs, a workable remedy was found. When, in due course, the engine was returned to English Electric in early 1958 a more permanent solution was sought, and much else beside.

Having acquitted herself fairly well, it was important to break the engine down into its constituent parts for examination, this was particularly so for the turbine and transmission. And when complete Hughes had a vast bank of material to consider:

All the information obtained during this exercise and from the tests at Rugby was then

used to prepare a detailed, written specification for the complete locomotive. From this a job list was created. This was partly an instruction on the nature of the new designs to be prepared and partly a list of modifications to those elements of the design highlighted during the tests for development. Months of hard work lay ahead to complete this task and make sure that the locomotive would meet BR's stringent requirements. To do this all items of equipment had to be built and tested, including the carriage heating boiler which had not been included in the first build, individually and together. With only limited resources at my disposal, because so much was being absorbed by English Electric's diesel programme, this proved most challenging.

Perhaps more important was the effect the test results had on English Electric's senior managers, who had to consider whether to let the project proceed to completion and allocate even more funds for this to happen. It wasn't a foregone conclusion that they would and there were several anxious months of waiting before approval was finally given. This process may have been helped by an agreement with the British Transport Commission for GT3 to be tested on the mainline by British Railways when complete. It is clear from correspondence at the

(Above) **As GT3's** time at Rugby drew to a close, John Hughes (in the foreground) was happy to entertain parties of interested individuals or, in this case, a small group of school children. By this stage the temporary canvas sheet, which had been stretched over a metal frame covering the cab, had been replaced by a rather more resilient metal cover rivetted to the frame. (Right) On the back of this print Hughes has wtitten, 'just returned from Rugby and ready for final construction to begin'. (JH)

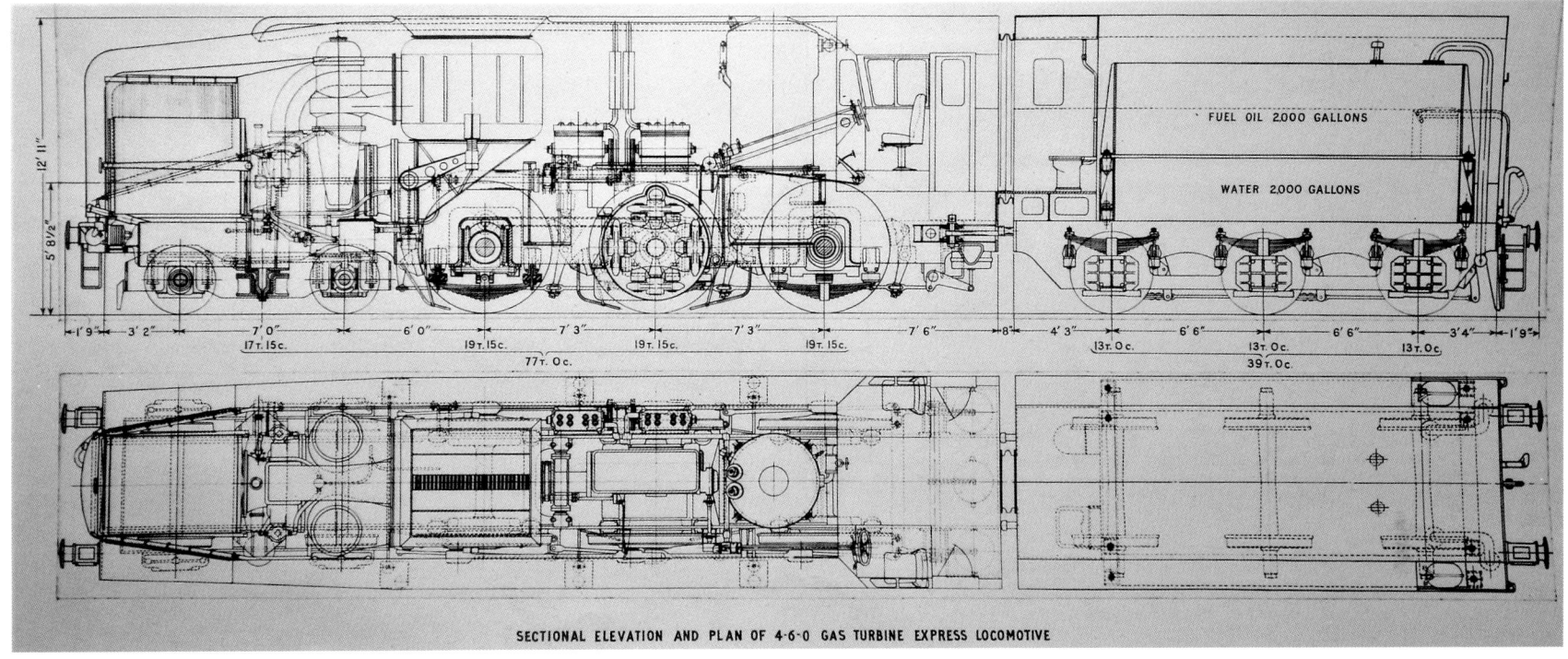

John Hughes gives no indication of the origins of this drawing. His only written comment was that it 'captures GT3 in her final form as turned out of the works ready for trials to begin in 1961'. (JH)

time that English Electric believed this suggested a genuine intention to purchase one or more of these engines, so bringing commercial success one step closer. Perhaps hope was triumphing over reality a little, but it was just one of a number of locomotive projects that were being successfully concluded at the time, under the auspices of BR's Modernisation Plan, so it was quite natural to assume GT3 would be accepted too.

Interestingly, in his surviving papers Hughes describes this final phase of the programme in general terms only and any comments he makes are brief and to the point. There are occasional references to the pace of the work and many photographs of the locomotive gradually being built,

but no word about any production difficulties or any additional modifications it was felt necessary to make. So, is it safe to assume that all went well and any problems that arose were easily dealt with? In the absence of any comment to the contrary, this is the only conclusion we can reach. In fact, Hughes's main focus seems to have been on the testing and evaluation process that would begin when GT3 was finally unleashed, with BR's permission, on to the mainline. He wanted to agree protocols for a testing programme that would:

> Establish that the general behaviour of the locomotive was satisfactory. In addition, allow BR, who would be handling

the engine at all times with our assistance, time in which to assess GT3 independently and confirm whether they would purchase it or not. This was the key to the project's continuation and success. To this end detailed negotiations with the British Transport Commission which established terms and conditions and a contract, of sorts, was agreed in January 1961. However, they would give no guarantee that they would then purchase the engine.

This was a great disappointment to me because I had hoped that our close co-operation over many years, plus the success of the trials at Rugby and transparency of our work at all stages of

Opposite and above: **Final fitting** out of GT3 was not a quick process, with progress being driven by the continued development and extensive testing of its component parts, following evaluation at Rugby. It wasn't until the latter part of 1960 that final assembly began. This group of photographs, which John Hughes collected over a twelve-month period, shows this work being completed. Of particular interest are the two pictures of its rather sleek tender being constructed. It was based on a standard BR six-wheeled chassis with a top specially shaped at the Vulcan Works to contain oil and water tanks, a train heating boiler, a crew compartment and a centrally mounted corridor On the back of the seventh print, Hughes has simply written, *'finally completed and running under its own power – a glorious sight after thirteen years of work'*. The relief expressed in these words is almost palpable and this was probably also felt by English Electric's senior managers who had kept the project going for so long. During December 1960, in celebration, they lined GT3 up with four of their successful locomotive projects for a publicity photo which was given a wide circulation (Above). (JH)

the programme, would have allowed BR to confirm purchase of at least one gas turbine engine. It was at this point I began to realise that all might not be well and final success might elude us.

Nevertheless, and despite this perceived reticence, they pressed on, now, perhaps, more in hope than certainty of commercial success.

Before gaining access to the mainline, BR insisted that its engineers be allowed to thoroughly examine and test the engine at the Vulcan Works to ensure that it was 'mechanically fit and safe to operate for a period of twelve months from the date the locomotive is delivered. After that BR will continue to use and operate the locomotive until this agreement is terminated by either party'.

To ensure that this proving period proceeded in a measured way the BTC agreed to set aside 'a suitable length of track in a siding for preliminary running trials and then provide for more advanced trials on branch lines', before letting the engine proceed onto the mainline. At the same time, the BTC selected an experienced driver, plus a reserve and a guard from Warrington Depot, who would be trained by English Electric to operate the locomotive and remain with it as long as the process lasted. And later on, they would be joined by other locomen from Crewe and Leicester so that sufficient cover for the trials, and normal day to day running, was available. However, in terms of maintenance, although workshop facilities were set aside by BR to maintain the engine, all work,

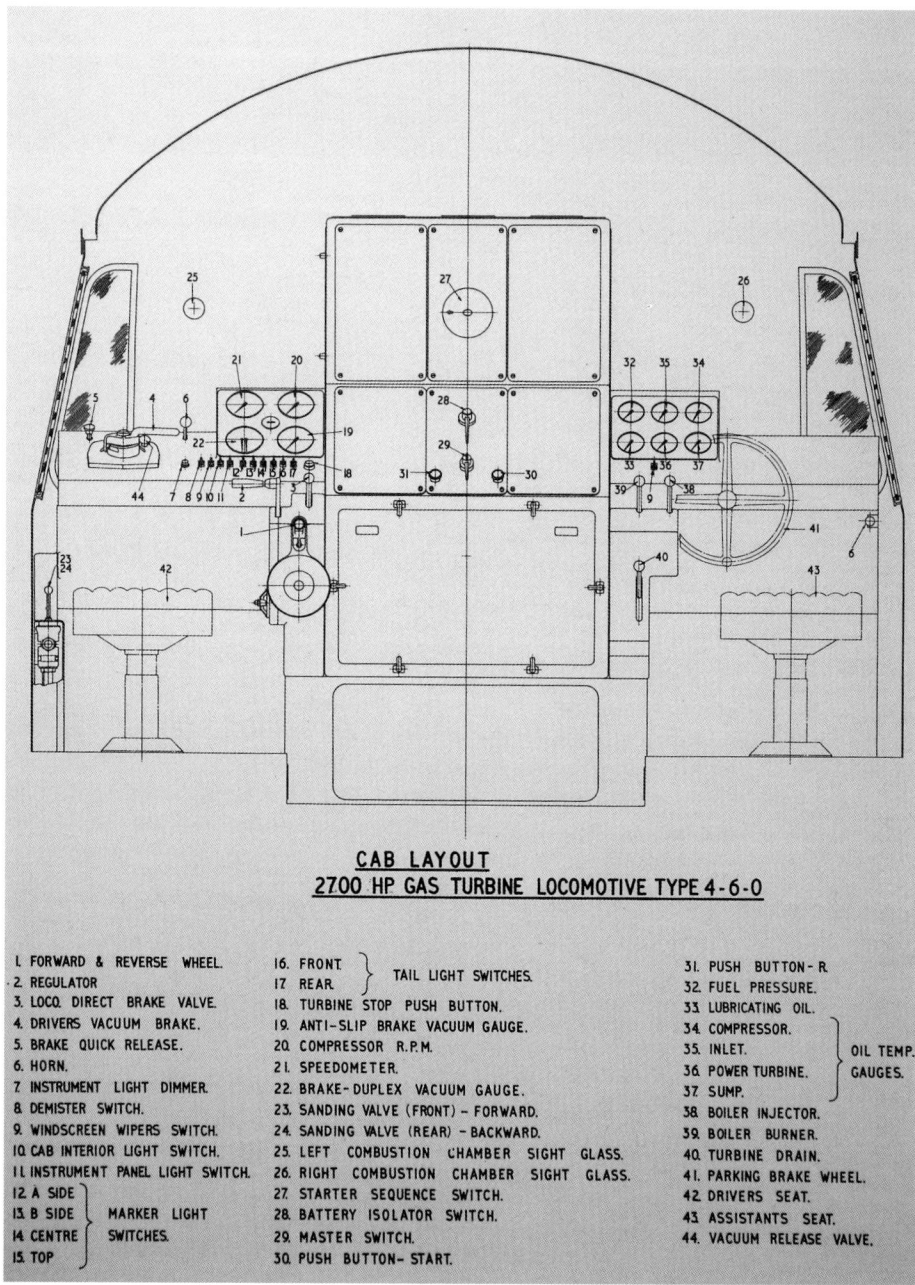

Above and opposite: To help BR driving and maintenance staff understand the technology that lay behind the development of GT3, before the locomotive was handed over to them English Electric produced a brief 'user manual'. These are just two of the many illustrations it contained – cab layout and the internal arrangement of machinery. When BR allocated particular members of staff to run the engine, they spent time with John Hughes and his team learning about its servicing needs and how it should be operated, so making transition, it was hoped, easier. (JH)

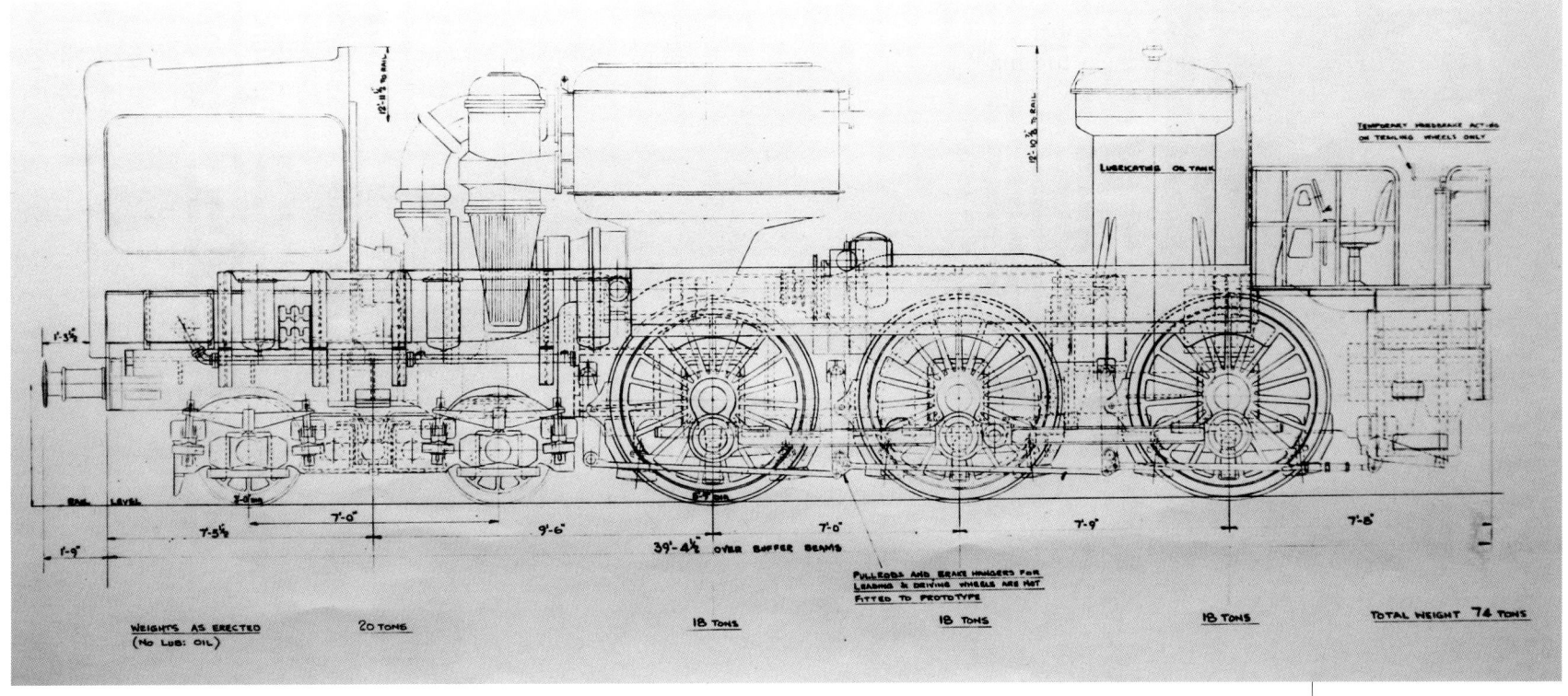

beyond daily servicing tasks, was the responsibility of Hughes and his team.

With agreement over the test programme reached in early 1961 training of BR staff to operate and maintain GT3 could begin. Their purpose, as Hughes recorded, was to 'establish that the general behaviour of the locomotive was satisfactory', though, interestingly, added no other defining parameters or targets to this very broad statement. With so much riding on the engine's performance in day-to-day service more might have been expected to have been written down and stated in public. But correspondence between the BTC and English Electric at this time is limited and what has survived suggests a growing lack of interest in GT3 by the customer and a weariness on the part of the company with the whole project. By 1961, English Electric were well established as a major supplier of diesels to BR, with DP1, having been transformed into a group of 22 Class 55 Deltics, leading the way, so there was an ever reducing interest in gas turbine locomotives. Coupled to this, this nationalised industry had already tested and rejected the GWR/Western Region Brown Boveri and Metrovick turbine engines. So, all in all, BR's future path had been set and yet another of the type seemed fairly certain to be rejected too, no matter how good it might be. Hughes probably sensed this, as did his masters, so the testing programme they now began seems to have been tinged with a pessimistic glow. But the commitment was made, work continued, for the time being anyway, with Hughes later recalling how the programme was rolled out:

After works tests the engine was towed to Whitchurch where, with the assistance of the London, Midland Region of BR, arrangements had been made for the first runs to be carried out on a little used branch line in Shropshire. Here faults could be brought to light and rectified without interference to traffic movements. One road to the disused engine shed was brought back into use as a base for these operations.

The engine was first run on the branch line by itself and then with coaches. During this period various defects in the reversing system came to light. These were mainly associated with the performance of the power turbine brake and the balance pistons. All this caused much delay in the testing programme. In addition, there was also spasmodic trouble with the starting of the turbine which was finally traced to a change in the setting of the spill pressure reducing valve due to a small internal leak. The other troubles that arose were minor by comparison and did not unduly interrupt the programme.

It was at this stage that the decision was taken, by BR and English Electric, to allow GT3 to join the group of locomotives selected to be part of the Institution of Locomotive Engineer's Golden Jubilee Celebrations at Marylebone. With BR dithering over whether to buy the engine or not, this does seem

(Opposite) **GT3 photographed** shortly after completion when visiting Whetstone where John Hughes and his team were based. (Above) In the early weeks of testing by BR GT3 was based in Shropshire where trials could be undertaken on branch lines where such tasks would not disrupt local trains services to any great extent. To allow this to happen, maintenance facilities were set aside, according to John Hughes, 'at Whitchurch, where English Electric's staff could undertake repairs and the engine could be serviced between turns. The arrangements there were quite basic and most work had to be carried out in the open where the elements made conditions quite challenging'. (JH)

a strange decision, for Hughes far less so. By this stage he was eager to promote the engine wherever and whenever he could. Such a public display, with the Duke of Edinburgh as its highest profile guest, would undoubtedly spark great media interest and this proved to be the case. So, in April 1961, the engine was detached from its duties and travelled to London under its own power. And in her unique brown paint scheme, highly polished for the occasion, she made an interesting sight eventually sandwiched between a Western Region Hymek diesel and a BR Doncaster designed and built AL5 electric locomotive. To say she stood out in the distinguished company of famous railway locomotives from the past and the newest products of BR and its suppliers would be something of an understatement. As one journalist wrote, 'this engine was most noticeable because of her unusual shape and colour – not quite a steam engine and not like any of the new diesels or electric locomotives on display. Prince Phillip lingered around her far longer than any other engine on display even the record breaking *Mallard.*'

If publicity can truly unlock doors that might otherwise stay shut, then the extent of press coverage GT3 received during 1961 should alone have ensured success. But life is never that simple and hard-headed officials in BR still remained to be convinced of the projects value when English Electric handed their locomotive to them for evaluation. John Hughes later described how this process slowly

The Golden Jubilee celebrations at Marylebone in May 1961 were probably the high spot in John Hughes's work on GT3. Here was a completed, fully functioning gas turbine locomotive finally about to make its mark on the mainline, albeit under trials conditions. The attention she received during those few days was immense, so Hughes's hope for the future must have been riding high as the great and the good of the railway industry, other VIPs and the press, clambered over her. Sadly, this optimism was not to last. (Above left) GT3's unnamed driver (but thought to be Driver Bebbington who was one of the men selected by BR to work GT3) and a BR Inspector pose beside their engine for Hughes. (Lower right) Inche Sardon bin Haji Jubir, the Malaysian Transport Minister, shows some interest. (Above right) And the Duke of Edinburgh, accompanied by John Hughes, takes time in a busy schedule to admire the new locomotive. Perhaps the guest most revered by Hughes, as a man, engineer and advocate of turbine power, William Stanier stands looking on (his head and white hair can be seen in the right foreground). (JH)

A photo thought to have been taken, according to John Hughes, at the end of the exhibition when the locomotives were being returned to their parent regions. The contrast between GT3 the standard class steam locomotive to its left, one of the first Class 55 Deltics and a member of the Southern Region's twenty-four Class 71 electric locomotives, is an interesting one. In shape, GT3 was clearly a halfway house between steam and the new generation of locomotives then coming into service. (JH)

evolved, though when writing in late 1961 many more months of testing lay ahead:

The stage had now been reached where longer runs with higher loads and speeds were required to establish that the locomotive would continue to operate reliably. It was agreed that the first aim should be ten consecutive journeys of 200 miles or more without trouble arising that would require attention being given to the locomotive between trips.

As a result, test runs took place between Whitchurch and Llandudno Junction with a load of 12 coaches to normal express timings. It had been agreed during negotiations in late 1960 that GT3 would undertake duties over routes normally assigned to Class 6P and 7P 4-6-0 ex-LMS locomotives and mixed traffic Class 5 4-6-0s and this is what happened now. After five successful trips the series was interrupted by a recurrence of the starting problems. Once repaired, three further trips were run, the final one with the load increased to sixteen coaches. The distance run was now 1,722 miles and a maximum speed of 72mph achieved.

The locomotive now entered into a period of daily running, working a test train twice a day between Leicester and Woodford Halse, South of Daventry in Northamptonshire, while other drivers were trained. After satisfactory running for three weeks it was decided to commence the required series of trials between Leicester and Marylebone. However, on the first run there was a recurrence of the power turbine brake slipping and so it was decided to modify the system before further runs took place. When this work was completed, eight more runs were made without any trouble occurring, followed by another five after a minor problem caused a postponement.

As there was now a satisfactory record of hard-running on this hilly route, it was decided to move the locomotive to Crewe for testing between there and Carlisle. The load of ten coaches on the first day was increased to twelve on the second and fifteen on the third. During these tests the locomotive showed itself the master of this load over this difficult route and thereafter eight consecutive trips were worked with twelve coach trains."

It was during this stage of the trials that a dynamometer car was added to the train for the first time at John Hughes's suggestion and with this the whole process of evaluation seemed to take on a more serious note. This suggests that BR's engineers had been closely monitoring the trials and were undecided, unlike John Hughes, about the progress being made.

During 1961, there was extensive coverage in the press about GT3, mostly promoted by English Electric's own PR Department. News items were carefully fed to newspaper editors, both local and national, and newsreel cameras were soon recording live pictures of the locomotive in operation to be shown at local cinemas. All this gave the engine a very high profile in the public eye and a response very reminiscent of Nigel Gresley's experimental 4-6-4 W1 engine that appeared in 1929 which was also given the nickname 'Hush Hush' in recognition of the secrecy surrounding its development. (JH)

The hush-hush loco
IT EVEN HAS BEDROOM FOR RELIEF DRIVER

AIR FILTERS | 2,700 H.P. GAS TURBINE ENGINE | CONVENTIONAL 4-6-0 STEAM LOCO CHASSIS | REST ROOM WITH BED | DRIVER'S CAB | FUEL TENDER

THIS is Britain's latest gas-turbine locomotive now having secret trials on the Whitchurch-Chester line in Cheshire, which has been closed to ordinary traffic. Its designer is Mr. John Hughes, who modestly says that there are "one or two new ideas being built into it." Note the huge air filters at the front for the turbine blades. And hidden in the back are a bed, wash-basin, and toilet for the relief driver. A 24-hour guard is being kept on this loco. The boffins working on it sleep in a local hotel and take their meals in the station buffet at Whitchurch. The loco was built at the Vulcan Foundry works of the English Electric Company at Newton-le-Willows, Lancashire.

By pursuing these more rigorously monitored tests he hoped, or so it seems, to force the issue towards a positive outcome or, perhaps, to an outright rejection of the engine. This would undoubtedly have been the worst news for him, but it would bring some certainty at least. The drip, drip, drip of uncertainty that had surrounded the project for so many years would finally, though disappointedly, be over and he could move on to other design tasks.

The dynamometer car was available in December and on the 13th and 14th of that month two days or trial running were planned. Sadly, the results, which were published in March 1962 (Report No. L163 which is reproduced in Appendix 3), did not make satisfactory reading for someone eagerly looking for an outcome that might take the project forward:

With the locomotive in its present condition, the power which is obtained at the drawbar with the driver's controller fully open, is subject to wide variations due to variable compressor speeds (7,500 to 8,250rpm). The variation increases with speed, and amounts to about 400hp at 50mph.

At certain times the compressor speed rose to its full value of 8,250rpm, but even on these occasions there was a serious deficiency in power

GT3 appears to be making light work of a heavy load during one of her trial runs between Crewe and Carlisle in late 1961. During a test on 14 December pulling 466 tons, her performance was deemed, by John Hughes, to compare favourably with a Stanier Princess Coronation and English Electric's new Class 55 Deltics. (RH)

output amounting to 330hp at 50mph.

In view of these undesirable features, it is considered that further dynamometer car tests would not be justified until the control system has been sufficiently developed to produce a reliable output."

Hughes added a little flesh to the bones of this slightly dismissive report, perhaps in a bid to broaden its horizons and make it more optimistic. So, in early 1962 he wrote this summary:

The power turbine system has proved itself during extensive static and mobile tests, at low and high speeds and when pulling heavy loads in difficult conditions. All other items of machinery have also, in some cases after modification, proved effective in service eg transmission, air intake system and carriage heating boiler.

It is significant that the most frequent causes of delay in the running of these tests were linked to defects in the power turbine brake system which for a variety of reasons was the least tested component at the start of track running. However, modification work during the course of this 'running in' period helped resolve these difficulties and allowed the engine to perform as effectively as it could in the circumstances that prevailed at the time.

GT3 has now reached the stage when it has run several thousand miles performing as its designers intended it should, and in all the track trials carried out during 1961 every

Below left and below right: **Throughout her trial** period with BR GT3 was occasionally captured by photographers – official or otherwise – with some of their pictures ending up with John Hughes, many, as in these two cases, without details of dates or locations. (JH)

return journey was completed if not always to schedule, at least without causing any delay to other traffic. On this basis it is safe to assume that the locomotive is capable of further development. However, many more thousands of miles running are now required to prove that its reliability and the life of its components are all that it is required in railway service and that the low maintenance, which is essential for success, is realised in practice.

With BR and John Hughes's reports to consider English Electric had to take stock of what they would do next. If they had been given any sign that an order might be forthcoming the decision might have been easier. But with development costs continuing to rise and no sign of a sale on the horizon, to BR or anyone else, although an expression of interest had been received from Poland, the laws of commercial economics would soon have to be applied. Quite simply, the company had to force a decision one way or the other. So, in March 1962, BR's General Manager, H.C. Johnson, was contacted and given a simple choice – buy and the project continues, don't buy and it will be brought to an end. He, in turn, wrote to John Harrison, BR's Chief Mechanical Engineer, on 12 March stating that 'Hughes has stated that GT3 will not run again until the firm have either orders or letters of intent to purchase'.

Over the next few days' discussions within BR seem to have resulted in a wait and see response and, in due course, the words 'we need to see more evidence of success before deciding whether to purchase or not' were communicated to the company. After so many years of investment they probably felt that English Electric would continue to play ball, but this proved not to be the case as Johnson again recorded, in very stark terms, in a second letter to Harrison on 20 March:

English Electric Co Ltd have decided not to proceed with

Two sides of English Electric's locomotive development programme of the 1950s and '60s. GT3 soon to become a back number rests besides an English Electric Type 3 diesel, soon to be designated a Class 37 in which guise 309 will be built between 1960 and 1965. (JH)

the running of this locomotive, but no doubt you will be advised of this directly by the firm in accordance with the relevant clause in the Terms and Conditions respecting the use and operation of this locomotive on British Railways.

Within days this decision was confirmed, with both sides being unable to find a compromise. During this period Sir George Nelson, English Electric's long standing chairman, had fallen victim to increasing ill-health and his influence had waned, as did his presence at work. To fill the void this created, his son, as Managing Director, had taken day to day control and, it seems, applied a tougher accounting regime to any speculative ventures that seemed unlikely to bear fruit. Ultimately, this meant an end to GT3 unless BR could be forced into a true commercial arrangement or if the interest from Poland was turned into an order. In both cases the answer was no and Nelson junior, who became chairman on the death of his father that July, put a block on this increasingly expensive project. This may have been a short term measure while markets were being tested, but with railways around the world choosing diesel or electric options, it soon became a permanent one.

Whilst these debates were taking place, GT3 sat in the Vulcan Foundry Works undergoing maintenance awaiting its fate. Meanwhile, John Hughes, who seems to have singlehandedly kept the project alive over so many years, was promoted to become Chief Mechanical Development Engineer of English Electric's Traction Division in 1962. This meant that he was no longer directly involved with GT3 and, in this new role, his influence over gas turbine developments was effectively brought to an end. From now on he would, amongst other things, be, 'investigating the case for reducing the number of traction motors used in electric and diesel-electric locomotives and the alternative of hydraulic transmission'.

Without its chief advocate and a sympathetric chairman to support the project GT3 was simply left to languish in English Electric's workshops. As she did so, Hughes, during April 1962, produced a project report which included some of the highlights of the testing programme undertaken with such hope. At the end he then described the problems that had been overcome along the way and in a covering note added a brief obituary:

> It is with some sadness and regret that I have to announce that this project is now coming to an end. I had hoped that GT3 would be built and lead to many of her number entering service with BR. We achieved the former but it seems the latter is now beyond our reach. However, there is much of which we can, justifiably, be proud, not least of all completing a prototype with limited resources and seeing it run successfully during static and mobile trials over many challenging routes in a variety of conditions. In the process it was agreed that the engine proved equal to Stanier's Princess Coronations and English Electric's own Deltics amongst others.
>
> All this happened at a time when the company were very heavily committed to BR's diesel and electrification programmes and, more broadly, to the highest profile engineering projects related to power generation, aviation and the production of domestic goods. In all we achieved we upheld the high standards by which the company has and will be judged.

Sadly, no other advocate came forward to keep GT3 alive and the company's interest in this engine was never revived. In the years that followed, the engine's carcase was slowly picked over with some parts, it seems, being used on other projects. With nothing more to be done with her, what remained was eventually sold to Thomas Ward Ltd of Salford and during the early months of 1966 she was finally scrapped.

As GT3's life was slowly brought to an end, there appeared in the *Railway Magazine* an article concerning the engine written by the noted railway historian O.S. Nock. As an independent assessment of the locomotive's worth it has great value, but as a hint of what might have been, if BR had had the courage to buy GT3, it is even more illuminating. As he did on many other occasions, Nock managed to join the locomotive under test. As an experrienced engineer himself, he was well-equipped to observe and assess its performance. In February 1962, his

According to his records, John Hughes exchanged letters with the railway historian O.S. Nock for a time in 1961/62. This resulted in Nock being invited to join GT3 for a test run from Crewe to Carlisle late in 1961 which resulted in a very positive article appearing in the *Railway Magazine*. Hughes sent Nock a number of photos, which were later returned. Two of these are shown here annotated by Hughes with the words (Left) 'GT3 passing over Dillicar Troughs before climbing Shap Summit' and ' our old faithful BR driver Mr Bebbington, who knew how to get the best out of the engine'. (JH)

report on GT3 appeared, a copy having been sent to John Hughes for comment late the previous year, before publication. In this he wrote:

On the day I travelled GT3 had a load of 12 coaches, weighing 384 tons tare behind the tender. The latter is corridor fitted and I was able to spend practically the whole journey on the footplate. There in addition to Driver Bebbington and the testing staff I met an old friend in Locomotive Inspector Drury, with whom I have previously enjoyed many miles of footplate travel.

Before beginning my evaluation, I must remark on the novel sensation of riding on a locomotive that was to all outward appearance a 4-6-0, but on which the 'firebox front' was replaced by a plain steel casing, finished in ultra-modern style in light mottled grey, and on which the corridor led through the middle of the tender with oil fuel tanks on each side.

We got away smartly from Crewe, and on the gently falling gradient towards Weaver Junction the controls were adjusted to keep the speed almost exactly at 70mph. But a slight signal check approaching Warrington proved the forerunner of a succession of signal stops, so much so that we lost all but 25 minutes on our scheduled 13 minutes from Warrington to Wigan. Two further checks due to engineering restrictions, and a momentary stop from adverse signals put us exactly 29 minutes down on passing Preston. To this point we were running on 'Special Limit' timings, the schedule of 57 minutes from Crewe being 3 minutes faster than that of the Royal Scot in 1933-5.

North of Preston, however, save for one slight check from signals at Brock, we had an absolutely clear road, and the existence of a long slowing for track repairs between Milnthorpe and Hincaster Junction provided an even more severe test of locomotive capacity in climbing Grayrigg.

The real interest of the run began at Carnforth, when from an initial speed of 72 mph the sharp rise of 2½ miles to Milepost 9½, inclined at 1 in 134, was cleared at 58 mph. Here and throughout the subsequent ascents, the locomotive was not being worked at full capacity. A slight leakage in the fuel system made it desirable not to extend the engine beyond seven eights full power.

The Milnthorpe permanent way check was an extremely long one, extending practically to Hincaster Junction and this latter point was passed at no more than 28 mph. Between Hincaster and Milepost 17 there are 1¾ miles of easier gradient – a mile of 1 in 193, and three quarters of a mile at 1 in 392 and here the speed recovered to 48 mph. After this a very gradual acceleration followed

GT3 at rest – date and location not recorded by John Hughes but her pristine state suggests she had either just been built or had just passed through a period of maintenance in the works. (JH)

until we were achieving 52 mph up the 1 in 131 on the high embankment beyond Hay Fell. This was far in advance of anything I have ever recorded with a Royal Scot. With GT3 there was a drop to 49 mph at Grayrigg Summit. This splendid climb involved an output of approximately 1,600 equivalent drawbar horsepower.

It must be added that the conditions were very favourable. Although cold, the weather was calm and fine, in total contrast to the previous day when most of the northbound run had been made in torrential rain. From Grayrigg speed was worked up in readiness for the climb to Shap; Tebay was passed at 77 mph and a brilliant ascent followed, in 6 minutes and 11 seconds for the 5.5 miles. Speed was still falling appreciably

On the back of this print John hughes has written, probably with some sadness, 'GT3 stored in the corner of the Works [presumably Vulcan Foundry] in June 1963 having been stripped of many parts her future uncertain'. (JH)

SUPER TRAIN'S LAST STOP

BREAKER'S YARD workers at Thomas W Ward Ltd, Ducie Road, Salford, will write the last chapters in the glittering life of locomotive GT3 this week.

The gas-turbine railway engine, developed by English Electric at their Newton-le-Willows works, will "die" in a blaze of acetylene glory.

The engine was planned as Britain's jet train of the future. A 15-strong team of engineers spent 12 years working on the top-secret loco, which became the most powerful single engine in Europe.

Built in 1961 as an experiment it proved to be very successful on its trial runs, but no one wanted to buy it because it was thought it would be uneconomical to run.

Pictured here with only a mileage of 11,104 the GT3 arrives at this scrap yard. Its turbines and instruments have been removed... but the crew's "mod cons" at the rear are still intact.

on the last mile, but even so, a minimum of over 40 mph with a 385 ton train was a most impressive performance, and registered a gain of no less than 7 minutes between Oxenholme and Shap Summit.

The locomotive was taken briskly down to Carlisle, though working a long way inside its maximum power, and when we turned off the mainline at Carlisle No. 13 Box the lateness had been reduced to less than 8 minutes. This concluded a most interesting and successful trip. My warmest thanks are due to the English Electric Company for the privilege of travelling with their engineers on this run.

To have impressed a writer of Nock's stature must have pleased John Hughes, but by the time the article appeared the death knell of the project had been announced so its positive content probably sounded a bitter note to a man who had been fighting to achieve his dream for so long.

Even though GT3's heyday in the public eye had long passed in 1966 the engine's demise still caused a small ripple in the press, as this brief article reveals. Up to this point, Thomas Ward Ltd had and would continue to be responsible for cutting up many of BR's withdrawn steam locomotives. (JH)

Epilogue
GT3 – THE UNREALISED DREAM

John Hughes's disappointment at seeing the project he had nurtured for so long quietly dropped, when success seemed only a hair's breadth away, can only be guessed at. The blow may have been softened by promotion to Chief Engineer in another division, but to have all his design ambitions thwarted in such a way must have been a heavy blow indeed. Worst still, he was unlikely to ever be in such a position again, the work he was then doing being, as he recalled, 'of a general management nature with little original thought required by comparison to my time with GT3'. Instead, his time was taken up 'reviewing other people's work and recommending its acceptance or not'.

A photograph John Hughes apparently kept framed and on display until the end of his life, according to his wife Felicity – a pleasant reminder of a project that absorbed him for many years and cutting-edge design work he doesn't seem to have been involved in again. In this case, he is seen posing with Roland Bond who, from 1953 to 1958, was BR's Chief Mechanical Engineer and then became the British Transport Commission's Technical Advisor. In 1960 Bond was visiting the Vulcan Works to assess, amongst other things, the progress being made on GT3. (JH)

For such a creative design engineer, who enjoyed the challenge of research and development, this must have seemed an ignoble and frustrating period in his career. And yet he 'continued to dabble in new ideas and invention whenever I could'. This included a proposed improvement to an engine cooling system by using a variable speed fan drive of his own invention and applying hydrostatic drives to locomotive auxiliaries. But, as he later wrote, 'this was not the same as designing, building and testing a complete machine'. So, when he was approached by the Managing Director of Associated Electrical Industries in 1964 to become the company's Chief Engineer, he accepted the post with some alacrity and left English Electric for what he firmly believed to be a return to the work he enjoyed best.

Two years later, perhaps frustrated by the limited opportunities as a design engineer with AEI, he decided to go into a partnership and formed Humphris and Hughes Consulting Engineers. But this also to failed to stimulate him as GT3 had once done. Then in 1967 he joined the Board of Motor Rail Ltd as their Technical Director. From here he saw out his career involved in various engineering projects, but none, it seems, gave him the same buzz as his years working on GT3. A semi-retirement to Cornwall followed and on 10 January 1977 he died when only 63 while living in St Buryan, five miles west of Penzance.

Judging by the size of the archive of papers and drawings he kept, it is probably safe to assume that GT3 was and remained of the greatest importance to him. But it is not complete because very detailed engineering drawings that underpinned the creative process and then allowed workshops to accomplish their tasks, do not seem to have survived or have not come to light yet. Nevertheless, such is the nature of this collection that we are able to follow many of his thought processes, gauge the intensity of his commitment and draw some conclusions. Perhaps, more importantly, for the eventual success or failure of the project, we can also follow the politics involved and the extent of his prompting that kept the programme going for so long.

From this it is clear that, without his drive and determination, the scheme may have remained a concept explored on paper only. But having taken it so far, and seen it become a living object, one can appreciate the deep sense of loss he must have felt, perhaps even a sense of betrayal, when it was finally cancelled. Yet he did get within an ace of success. If, in early 1962, John Harrison, as BR's CME, had indicated a willingness to buy the prototype at least, English Electric would probably have allowed work to continue for at least another year and seen the locomotive working normally across the network as intended. And who knows, this might have led to more being built for use in Britain and even encouraged an overseas market to grow.

Sadly, much of the correspondence surrounding English Electric and BR's decision-making process has not survived. But, as luck would have it, between February and April 1962, when GT3s future was still in the balance, Hughes presented his paper entitled 'The Design and Development of a Gas Turbine Locomotive' to meetings of the Institution of Locomotive Engineers, the last of these in London. By chance, this session was chaired by Harrison, the Institution's President that year, and was attended by other senior BR managers, some now retired, who had been involved in the GT3 programme. The most important of these was Roland Bond, who was one of Harrison's predecessors as BR's CME and President. Together, in the question and answer session that followed the presentation, both men, and many others, passed a mild form of judgement on Hughes's work. However, Harrison, perhaps aware of his management role in whether to order a turbine locomotive or not from English Electric, demonstrated diplomacy, perhaps allowing others to express stronger views on his behalf:

> All will agree with me when I say that Mr Hughes is clearly an enthusiast for his work. He has spent many years developing the gas turbine locomotive step by step….. It is most gratifying to listen to someone telling the story of the development of such a locomotive, and to realise the time and energy which he has spent in getting it to perfection."

Bond, who could be more open in his comments, no longer playing any active part in the selection process, then added a slightly more critical assessment of the project:

> This is by any standards an outstanding piece of engineering design and development.

At the time that decisions were being taken ideas were different to what they are today. The decision, for example, as to the general form the locomotive should take. Was it to be a double-bogie, double ended locomotive? Should it be, on the contrary, a 4-6-0 locomotive of conventional mechanical design? The decision then taken was, in the speaker's opinion, right. Today it would have been different.

As the author has pointed out, the locomotive has so far run about 11,000miles. Clearly, one could not yet reach a final verdict based on the criteria which must be used: ultimate operating and maintenance costs, reliability and so on. The locomotive must run longer to enable a clear indication to be obtained of these important points. Even though the locomotive might be too late in the change from steam to make an impact on motive power policy in this country, one hoped very much that it would be possible for this very notable piece of locomotive engineering development to be continued further so that it can be judged at its true value.

Dennis Carling, who had been in charge at the Rugby Testing Station when GT3 had undergone trials

With GT3 now a living machine those who had built her were quick to advertise the engine's existence (Left) and soon took and circulated many photos (Below) some of which were soon published in the railway and national press. (JH)

there, then added a very telling comment:

> I think that there are a number of railways abroad where such a locomotive would have a very fair chance of success. In particular, these would be any lines where operating methods and traffic requirements permitted the locomotive to be suitably loaded so that its fuel costs per unit of work would compare well with that of other types of locomotives. And, in addition, where its inherent characteristics could be made use of.
>
> It should be noted that for sustained power output at the drawbar this locomotive is only outclassed by one type, other than straight electrics in this country. It has the power to move traffic but only further running would show if it has the necessary stamina. Personally, I think that it would have, and hope it will be given the chance to show it."

So here, in an unofficial way, is BR's response at a time when English Electric were considering whether to proceed with the project. In essence, they needed more time and considerably more live running to decide whether to place an order or not, with Carling adding a very telling comment about the alternative possibilities inherent in the overseas market. Was this a subliminal message, perhaps? BR cannot or would not accept the engine, but someone overseas might be prepared to take the gamble!

Amongst all the other comments made by members at these sessions praise for Hughes personally were universally expressed, but they zeroed in on two crucial and critical issues – the length of time English Electric had taken to develop the engine and the choices made from the outset. The most telling of these came from B.T. Scales, R.J. Jarvis (a senior BR manager) and E.R.M. Montague:

> I am a little disturbed at this development taking thirteen years.
>
> I think that after that passage of time it was inevitable that the locomotive appears, perhaps, a little outmoded in having this particular wheel formation.

Whatever the reason this locomotive [4-6-0] type was chosen, for this most important development work, it was virtually an obsolete design from the first. The need for turntables at terminal stations and the poor visibility provided by the arrangement chosen all tended to prejudice the locomotive whatever technical merit the prime mover and transmission might have had.

Hughes's response to the comments of such influential and knowledgeable men was sanguine and defensive at the same time. So, perhaps, he still hoped that English Electric's managers might, even at this late date, be persuaded to change their minds and allow the project to continue a little longer:

> Had its withdrawal for reasons of economy been foreseen, then the running programme carried out in 11 months of 1961 would have been conducted differently, and minor troubles would have been lived with in order to run a bigger mileage. As it was the work was conducted with the conviction that all defects brought to light should be remedied to achieve the best possible record on entry into regular service.
>
> A project such as this locomotive requires a sustained effort by the organisation concerned with its design and construction and a belief in its future by a possible user to carry it forward.... Viewed in this light, the time given to track running was quite inadequate and not in proportion to the effort taken to get the locomotive into being and on the rails, and as has been pointed out, this left many questions unanswered.

So, a critical point had been reached in the life of GT3 with English Electric seemingly using BR's reticence to bring about its end, even though their reasons for not wishing to order a turbine locomotive just yet were valid ones. But commercial imperatives, plus BR's very public commitment to diesel and electric locomotives, could not be ignored by the company. And, in truth, Hughes had been given far longer than was strictly necessary to prove his ideas could work. Then, of course, there was the growing realisation that GT3 had simply been built to the wrong configuration. If it had been completed in the early 1950s, when steam still dominated the railways, things might have been different, but by 1960 the 4-6-0 concept was simply out of date. With hindsight it might have been better if the alternative layouts Hughes and Bill Allen had explored in the 1940s had found favour. Yet even then BR might not have been persuaded to go down the gas turbine route, having already rejected the models produced by Brown Boveri and Metrovick for the Western Region. And these had both adopted a modern twin cab, twin bogie configuration.

In some ways, having invested so heavily in its creation, it is a shame that English Electric didn't keep the completed engine themselves and use it as a means of demonstrating their turbine work. In an industrial form it was still being developed and sold worldwide, so why not use it for advertising purposes especially when it had already attracted a great deal of publicity. And from there it might have been preserved for posterity like the prototype Deltic. But someone, possibly the new Managing Director, thought otherwise and chose otherwise and this valuable experiment was allowed to languish and then be scrapped in what some thought to be an act of scientific and cultural vandalism.

So ended a story of great endeavour in which tenacity and a clear, and possibly better vision of Britain's railway future nearly succeeded only to be beaten, after a long battle, by the time it took to build, the outdated nature of model they chose to adopt and the lack of an alternative fall-back option. For English Electric this was effectively an end to developing a gas turbine locomotive but others in Britain had not quite given up yet.

John Hughes has written on the back of this photo 'my team and the BR driver service GT3 at Carlisle Shed before heading south again with another heavy load. While at Carlisle an oil leak had to be repaired, but apart from this GT3 was in fine fettle'. (JH)

In the months leading up to a very public announcement in November 1966, BR planners had been considering how they might improve the mainline, high speed passenger services. Since the war, these services had been haemorrhaging huge numbers of passengers as the switch over to cars reached record levels. In taking this project forward a partnership with Rolls Royce was established

Before the gas turbine powered APT(E)'s design had been finalised BR's engineers began preparing possible outlines for model makers to create in 3D form. This photograph captures one of their early efforts. Unlike GT3 they didn't make the mistake of adopting an obsolescent cab and wheel configuration. (JH)

and looked first of all at modifying existing multiple unit stock used on the Edinburgh to Glasgow line by fitting them with 'with small gas turbines of about 400hp designed by Pratt and Whitney'.

As the study progressed under the guiding hand of BR's Dr Sidney Jones, a 'radar and infrared specialist' who had led in the development of the Firestreak missile system, then headed the armament department of the Royal Aircraft Establishment before joining British Railways, other possibilities were explored. Most important amongst these was an experimental high-speed turbo-jet powered train. This proposal hit the headlines in July 1967 when its potential speed of 150mph became known. With the pre-war record-breaking feats of Gresley's A4 Pacifics still fairly fresh in the mind, this new train caught the public's imagination. But the gap between a proposal being suggested and the appearance of such a train was a vast one on cost grounds alone, especially when the need to upgrade parts of the network to allow such fast running to take place were taken into account.

Nevertheless, this proposal eventually gained acceptance and the Advanced Passenger Train (APT) project was born, with work beginning in earnest during mid-1969. The initial plan was to build an experimental tilting train with two carriages at either end containing 1,500hp Rolls Royce Dart gas turbine engines and two passenger carriages sitting between them. These middle cars would be fitted with measuring and recording equipment, so that all aspects of operation could be carefully monitored and assessed. As work progressed Dr Jones became aware

GT3 – The Unrealised Dream • 143

of a less expensive, but effective turbine engine being by Leyland for its truck division and the Dart option was soon dropped. As a result, power would now be supplied by four 300hp turbines in each power car, with a fifth turbine to meet the needs of the 'passenger' carriages.

In the two years that followed, this work progressed fairly quickly, unlike GT3, and in late 1971 a test train, given the title APT-E(Experimental) was ready for evaluation and development, if necessary. As a result of this work, the turbine fitted to generate power delivery to the passenger carriages was modified, during 1974, to provide additional power to the traction motors. At the same time, it was decided to replace all of the turbines with an upgraded 330 horsepower version, which improved total power output per car from 1,200 to 1,650hp.

And so, trials went on until 1976 when the test train was finally withdrawn from service for preservation. However, by this time serious consideration had been given to an alternative source of power. With BR's West Coast electrification now well advanced this made sense, but there was also the question of Leyland's exit from the turbine market to consider. This company had discovered by then that turbine powered trucks were not economically viable, especially so with oil prices rising so rapidly after the 1973 Arab-Israeli War. By this time BR had already begun to develop an overhead supplied electric APT and now this option took precedence. So, finally, BR's flirtation with gas turbines came

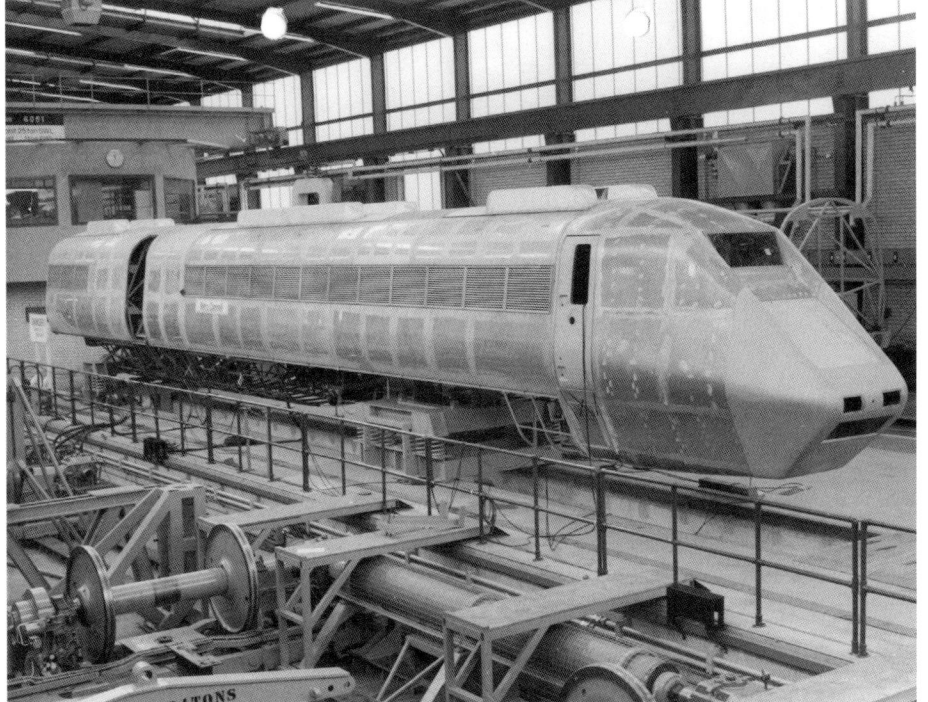

APT(E) takes shape within BR's workshops at BR's Advanced Projects Laboratory and is then unveiled to a waiting public before trials begin. Such a cutting-edge project continued to attract great attention, especially when the trials vehicle's successor was ready to begin service in late 1981. Unfortunately, during its first passenger carrying run on 7 December that year, it was considered to have fallen short of expectations, particularly its tilting mechanism, and a poor press ensured that searching questions were asked about the train's viability. This, ultimately, led to its cancellation. (JH)

Once again John Hughes kept no notes with this negative so we are left to speculate when in 1961, or even '62, it was taken. The presence of a GWR Pannier tank engine is suggestive of the Western Region. (JH)

to an end and APT(E) entered the record books alongside all the other turbine experiments dating back to Guiseppe Belluzzo's prototype engine of 1906.

Very little now remains of GT3 – photographs, reports, some cine film, drawings, adverts and so on – but that is all. And yet it is still remembered by those who witnessed its brief moment of glory in 1961 when, for just a few short months, it seemed possible that John Hughes's determination to build such a unique and quirky engine, against huge odds, might actually succeed. How he managed to push the project as far he did is quite remarkable, but at the last moment common sense prevailed, or so some believed, and this long running saga was brought to an end. But one is left to wonder what might have been if this hadn't happened and GT3, possibly accompanied by others of the class, had entered service. We shall never know, of course, but the prototype had begun to show herself capable of performing well and so all might have been well. Would this have made gas turbines more acceptable to BR and led to even more being built of different types? Probably not, because a policy of dieselisation and electrification had taken hold and other types of locomotion would only have been an unnecessary and possibly expensive distraction. So GT3 passed into history – gone but not forgotten.

Appendix 1
ENGINEERING AND AESTHETICS – HOW GT3 TOOK SHAPE

Long before John Hughes led the team that eventually built GT3 he had begun to speculate on the shape it might take. Normally a design engineer would consider function before form, but in this case Hughes, even before a suitable turbine power unit had been built, had begun to give much thought to the aesthetics of the locomotive he wished to design. In this process he was helped by Bill Allen, who, apart from anything else, was a talented artist and graphic designer. Whether he was amateur or professional is hard to determine now, but his work was

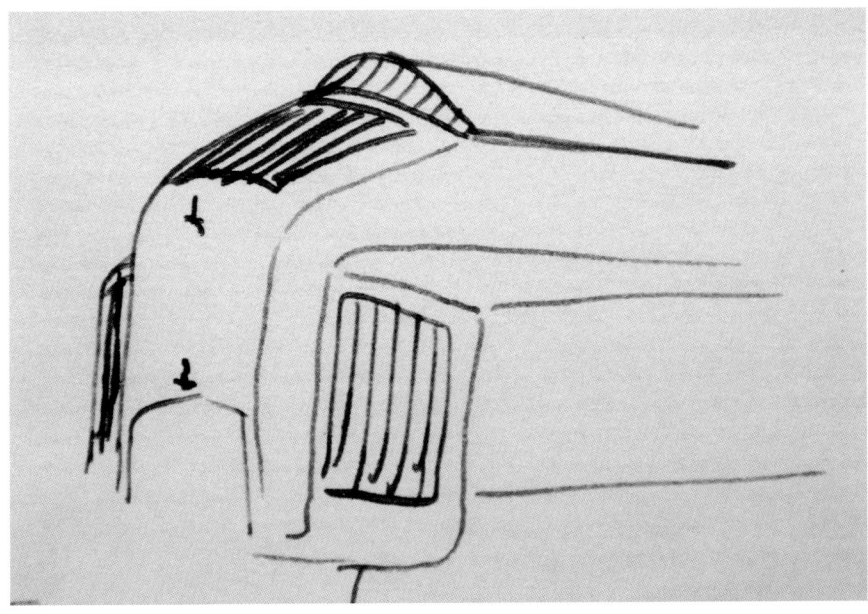 

As John Hughes contemplated plans to build a series of gas turbine engines, he enlisted the help of someone he appears to have known since the 1930s – W.(Bill) G. Allen. Over a period of twenty-five years or more they kept in touch, met frequently and corresponded on a wide range of railway related matters. They swapped ideas about external design so sparking ideas in each other to be explored further (as the two sketches above, which focus on the front end, make clear). This work gradually evolved into the exterior design of GT3. (JH/WA)

highly accomplished suggesting he had had some formal training. Either way, he and Hughes became friends and discovered a shared interest in railways locomotives and modelling.

As early as the mid-1930s they corresponded, with many of their letters surviving to provide a guide to what happened. It was through this developing relationship that Hughes was able to explore his ideas on form and in Allen he found a man who could translate his thoughts and some rough sketches into graphic illustrations of what might be, offering variations to reflect various wheel configurations or ideas adopted by other locomotive designers. However, Allen never went beyond this brief to provide drawings of internal layouts or offer any professional advice on purely engineering issues.

It is clear from what they wrote that both men were fascinated by locomotives and keenly observed all that was happening on the railways here and overseas as new models appeared. In so doing, they both appear to have devoured the contents of any technical journals and railway magazines they could find. And as these thoughts developed what they read clearly informed their emerging ideas. So such things as the types of wheels used on Bulleid's Pacifics, Gresley streamlining experiments and much more enter into their discussions. Then all this begins to influence the shape of their designs even before the internal structure has been considered. Some might say this is a classic case of putting the cart before the horse, but to my mind they were simply pursuing a creative dialectic. By adopting a brainstorming approach, they were able to generate ideas without practical constraints, then gradually focus these often theoretical concepts on ways of solving particular design issues possibly in a new and unique ways.

In the months and years that followed they considered many options, each with added variations as different ideas occurred to them and needed to be rehearsed to see if they held water. In this way they developed a variety of ideas for different forms of gas turbine engines and wheel arrangements, two of the most popular being 4-6-4 and 4-8-4 initially. These designs, which had cabs in the middle, were thought best suited to fast freight services, though why they chose such a unique arrangement isn't explained in the letters that have survived.

From here they moved on through other interesting options to consider a 4-6-0 design. As both men were clearly fascinated by steam locomotives the choice of this most famous of wheel configurations was, perhaps, understandable. But at the same time it might be said to be short sighted in ignoring what was happening in the world of diesel and electric locomotion, both of which were then receiving wide coverage in the railway press. Hughes would later write that going for a 4-6-0 configured gas turbine engine might make the transition from steam easier for those who find change difficult as well as providing the level of adhesion he required. At the same time, there was the issue of cost to consider. Building on existing, well-tried frames and wheels would be significantly cheaper than an entire new build. But in doing so, he was, some thought, saddling himself with an attractive but outdated solution. Nevertheless, their focus shifted to a 4-6-0 design, as English Electric seriously began to consider building such an engine, and in the years that followed they rarely strayed from this view.

During the time they corresponded, Allen's work for Hughes seems to have been largely voluntary in nature. Then, as English Electric gradually embraced the idea of building a gas turbine locomotive to showcase their EM-27 power unit, his design skills were rewarded with at least one paid commission. Because of this, the work he produced became more thorough and graphic in nature. Gone were the rough sketches, for the most part, to be replaced by highly skilled paintings and drawings of the proposed locomotive and possible derivatives. Much of this output was publicity driven – English Electric producing a number of brochures to send to prospective customers – and some was to help the engineering programme, where it would sit beside draughtsman produced technical drawings.

Undoubtedly, Allen's work helped Hughes focus his mind on various design issues relating to GT3 and, in so doing, acted as a valuable sounding board in a most creative way. Having someone independent to bounce ideas around with, allowing your imagination to roam in the process, can be truly invigorating and in this case appears to have been so. And for a decade, as the project gradually came to fruition, this relationship played a key part in the design process.

Luckily, because John Hughes was an archivist at heart, a considerable bank of information – reports, drawings, photographs, letters and paintings - has survived from this period. This in itself is most unusual, because BR were very careless with the historic records they, for the most part, inherited. This has created huge gaps in our understanding of the way locomotive projects were developed, so making the completeness of the Hughes archive doubly important. What follows is a small taste of the information he carefully collected, in this case focussing on the way one aspect of the design was discussed and slowly evolved. The drawings are all the work of Bill Allen working from his home in Belper and begin with some of the earliest sketches from 1948, ending when Hughes's efforts reached fruition and GT3 was born.

Left and below: **Often in** their correspondence Allen and Hughes would discuss wheel configurations for each type of gas turbine engine they thought might be needed. One of the more interesting of these was a locomotive with 4-8-4 arrangement which both men believed might be suitable for working high speed, heavy freight services. This concept, which first mentioned in correspondence during 1948, is described by Allen as having 'decoration that follows Great Western 19th century practice … a polished brass moulding, broken only by the loco name, which, I suggest, should consist of brass letters…'. The choice of name was an interesting one – Sir George Nelson was chairman of English Electric from 1933 to 1962. Suggesting that this might adorn the new locomotive had obvious benefits when and if the time came to seek board approval to build such an engine. (JH/WA)

ALTERNATIVE TREATMENT FOR 4-6-0 BASED ON "MERCHANT NAVY" STYLING.

Note: This style has been added in pencil to the scale layout.

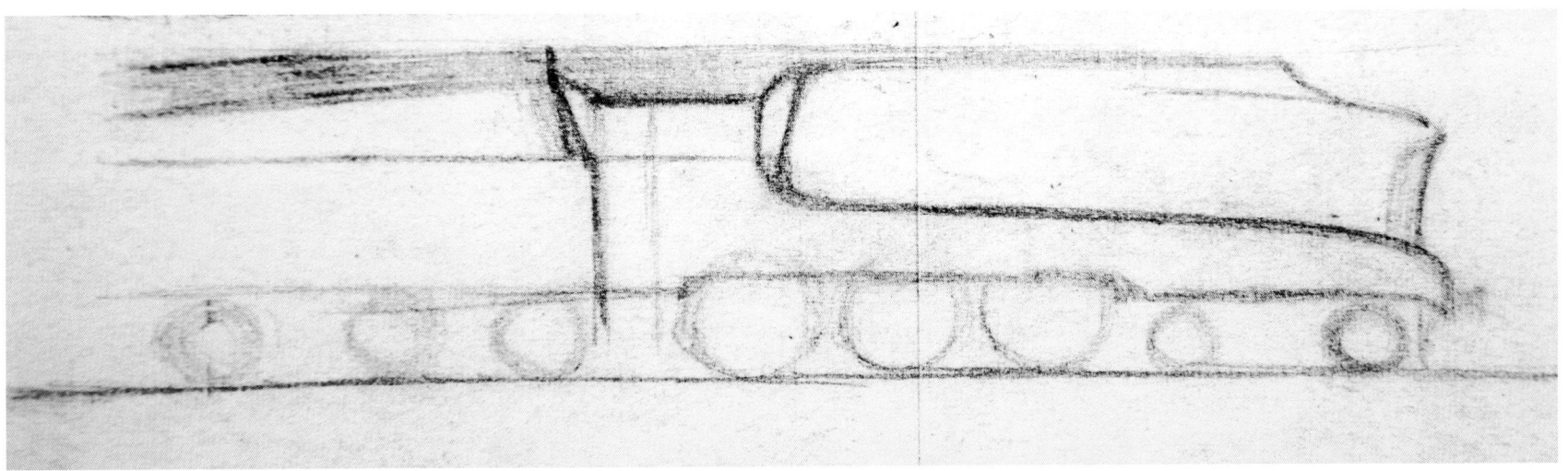

Opposite and above: **Both Hughes** and Allen were fascinated by steam locomotives and 4-6-0 engines in particular. As time passed, and English Electric moved towards approving development and construction of a prototype gas turbine engine, they came to believe that this wheel configuration best met their needs and over a six-year period their evolving designs rarely strayed from this specification. The drawings above are just three from a large collection showing some of the refinements they considered even before approval to build such an engine was given. (JH/WA)

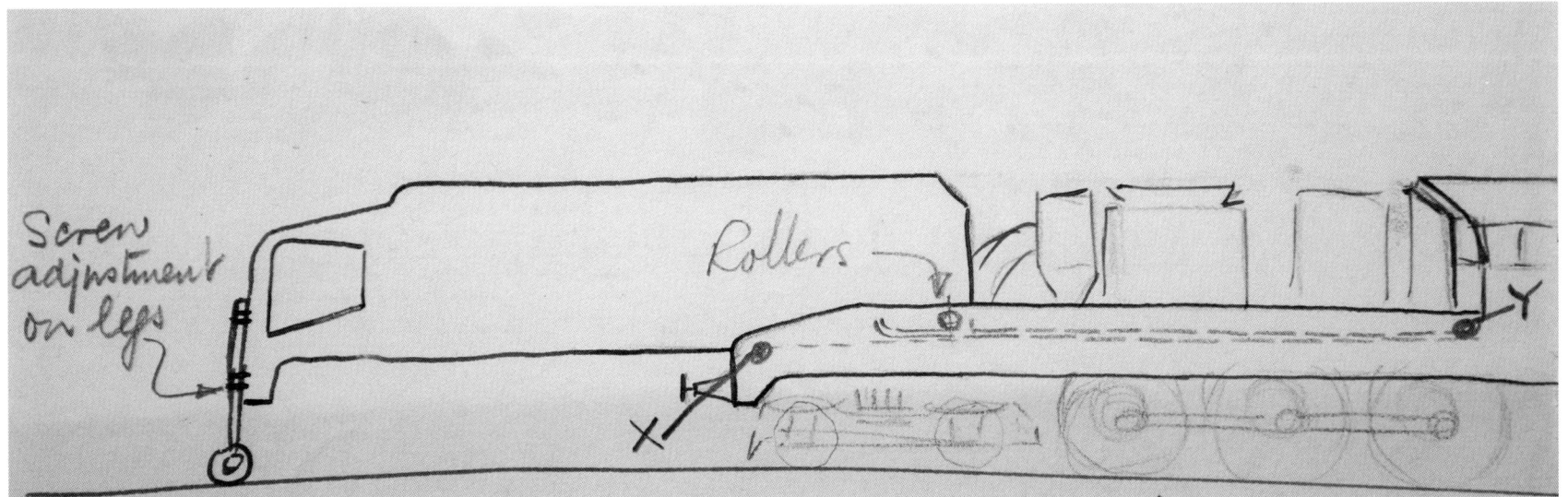

Once committed to a 4-6-0 design Bill Allen then moved onto the practicalities of such things as access for servicing. In due course he came up with this unique arrangement – a body that slid forwards. It was an idea that doesn't appear to have been developed further than this sketch, but at least it reveals an openness in considering new and novel solutions. (JH/WA)

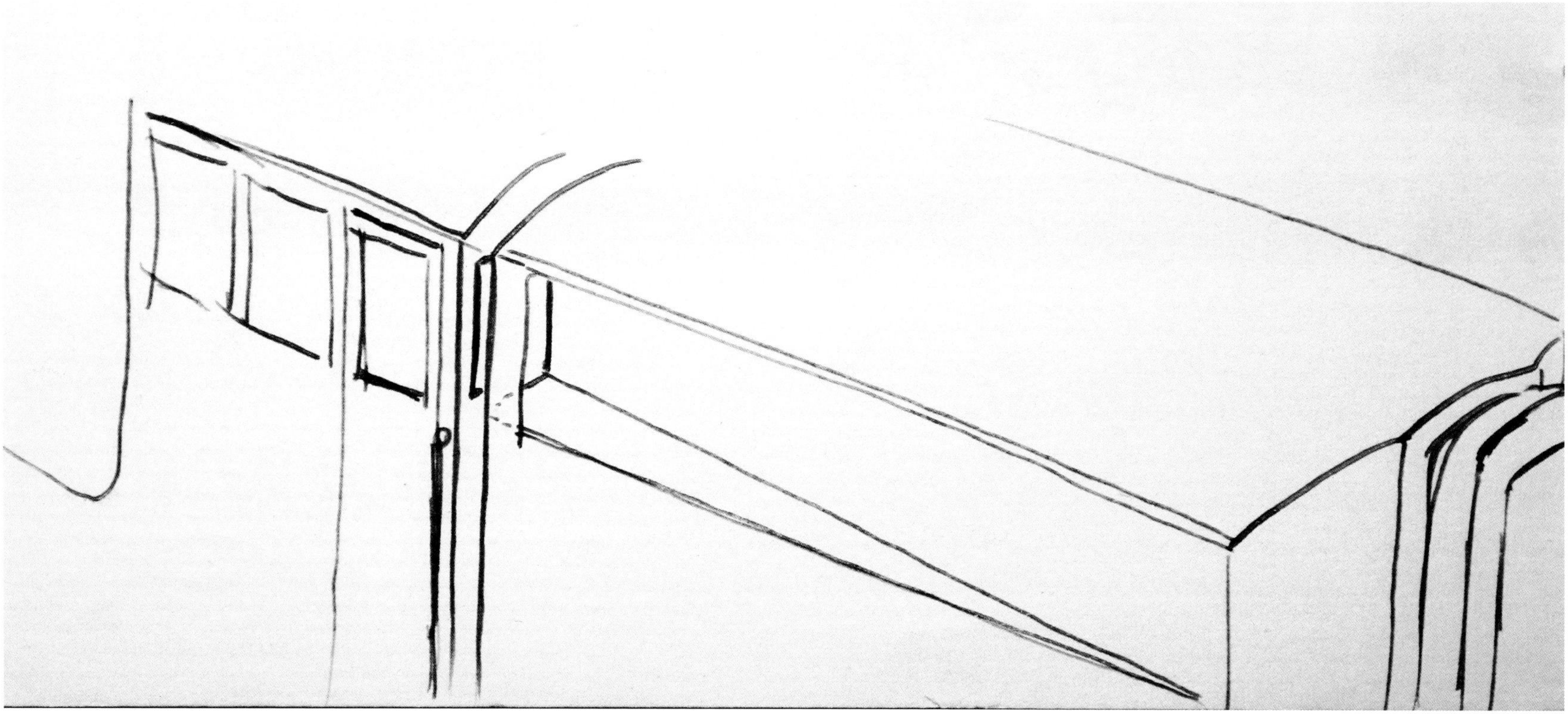

Having settled upon a 4-6-0 solution, though still keeping other options in mind, Hughes and Allen began to consider the fine detail of such a design, including ways of improving the view rearwards from the cab. In doing so, they were addressing the perennial problem steam locomotive designers had faced for many decades without the success so easily achieved by having a cab at the front, or cabs at either end. Bulleid had tried to do otherwise with his steam powered Leader Class in the post-war years, but with such a poor overall design success eluded him no matter where the cab was located. (JH/WA)

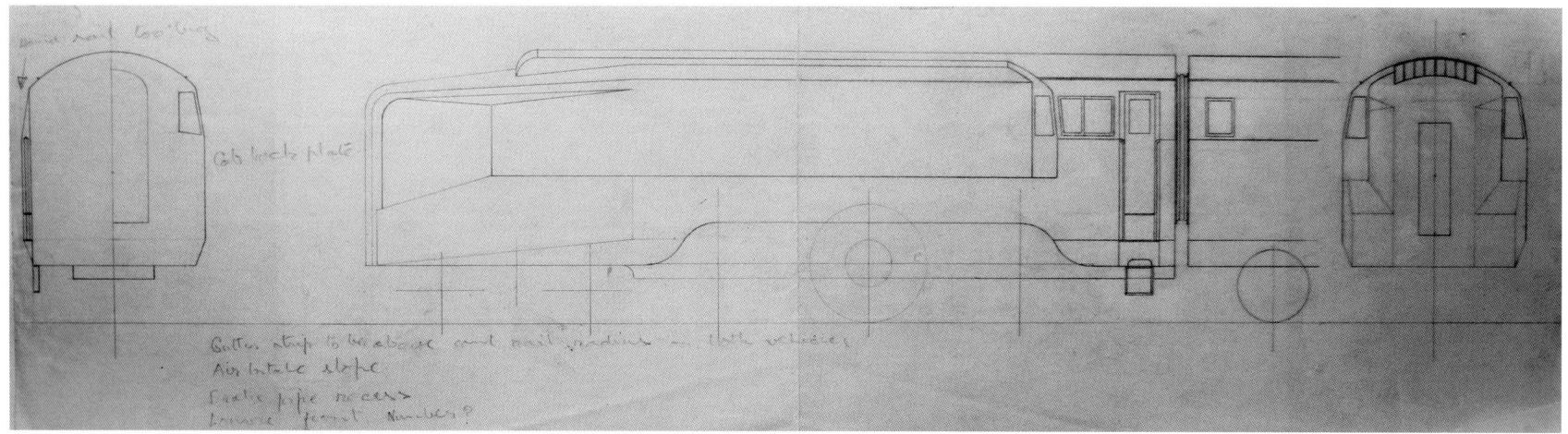

(Above) **And so** the evolution of ideas continued to flow with this outline drawing beginning to show how thoughts were coalescing around a shape that is remarkably close to GT3's eventual form. (Opposite above) Such was English Electric's confidence that such an engine would be built their draughtsmen soon produced a more detailed set of drawings, such as this shown here, adding much needed detail about its internal structure and how the various parts of the turbine might be fitted into the body. (Opposite below) Armed with this information, Allen produced this schematic which was widely used in publicity material sent to potential customers in Britain and abroad. (JH/WA)

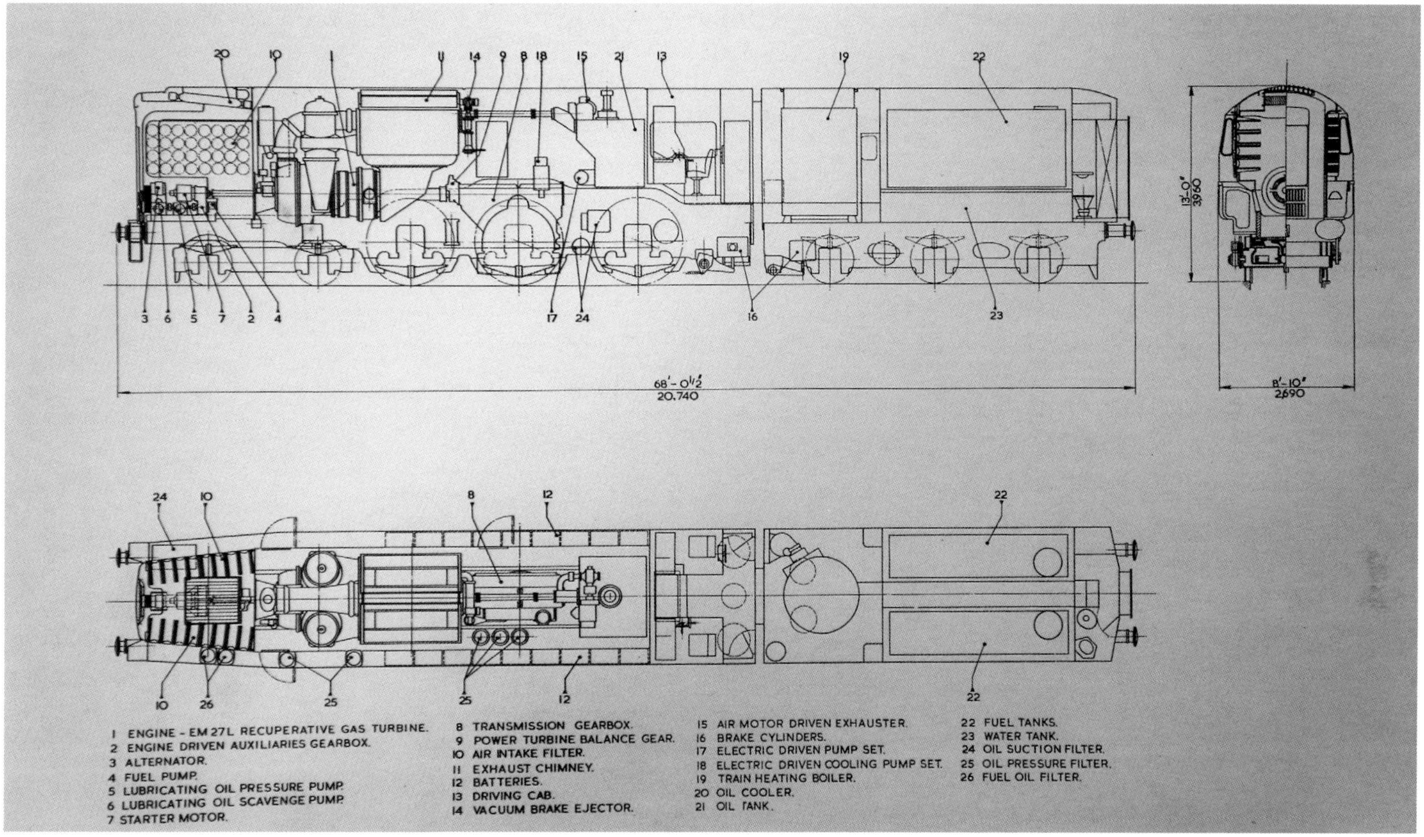

Engineering and Aesthetics – How GT3 Took Shape • 151

1 ENGINE – EM 27 L RECUPERATIVE GAS TURBINE.	8 TRANSMISSION GEARBOX.	15 AIR MOTOR DRIVEN EXHAUSTER.	22 FUEL TANKS.
2 ENGINE DRIVEN AUXILIARIES GEARBOX.	9 POWER TURBINE BALANCE GEAR.	16 BRAKE CYLINDERS.	23 WATER TANK.
3 ALTERNATOR.	10 AIR INTAKE FILTER.	17 ELECTRIC DRIVEN PUMP SET.	24 OIL SUCTION FILTER.
4 FUEL PUMP.	11 EXHAUST CHIMNEY.	18 ELECTRIC DRIVEN COOLING PUMP SET.	25 OIL PRESSURE FILTER.
5 LUBRICATING OIL PRESSURE PUMP.	12 BATTERIES.	19 TRAIN HEATING BOILER.	26 FUEL OIL FILTER.
6 LUBRICATING OIL SCAVENGE PUMP.	13 DRIVING CAB.	20 OIL COOLER.
7 STARTER MOTOR.	14 VACUUM BRAKE EJECTOR.	21 OIL TANK.

4.6.0. GAS TURBINE LOCOMOTIVE

During the 1940s some designers began investigating the benefits of using what were called Boxpok wheels. This concept saw a wheel being made in a number of box sections, it being believed that they would gain considerable strength from this method of construction and improve the stability of locomotives running at high speed. In Britain Oliver Bulleid was much taken with this idea and developed a British version to be used with his Pacifics, 0-6-0 Q1s and then the Leader Class. Here the wheels comprised of a single disc, with teardrop-shaped hollows to help give them rigidity, and reduce weight by up to 12 per cent. However, in service they proved difficult to maintain - the hollows collecting oil, sand and water. In addition, they were found to be prone to cracking, although one reason for this might have been the poorer quality of steel used during and after the Second World War. The use of the Boxpok wheels was discontinued in 1948. However, for a time they were given serious considered by John Hughes and Bill Allen, as this illustration of a gas turbine locomotive makes clear. (JH/WA)

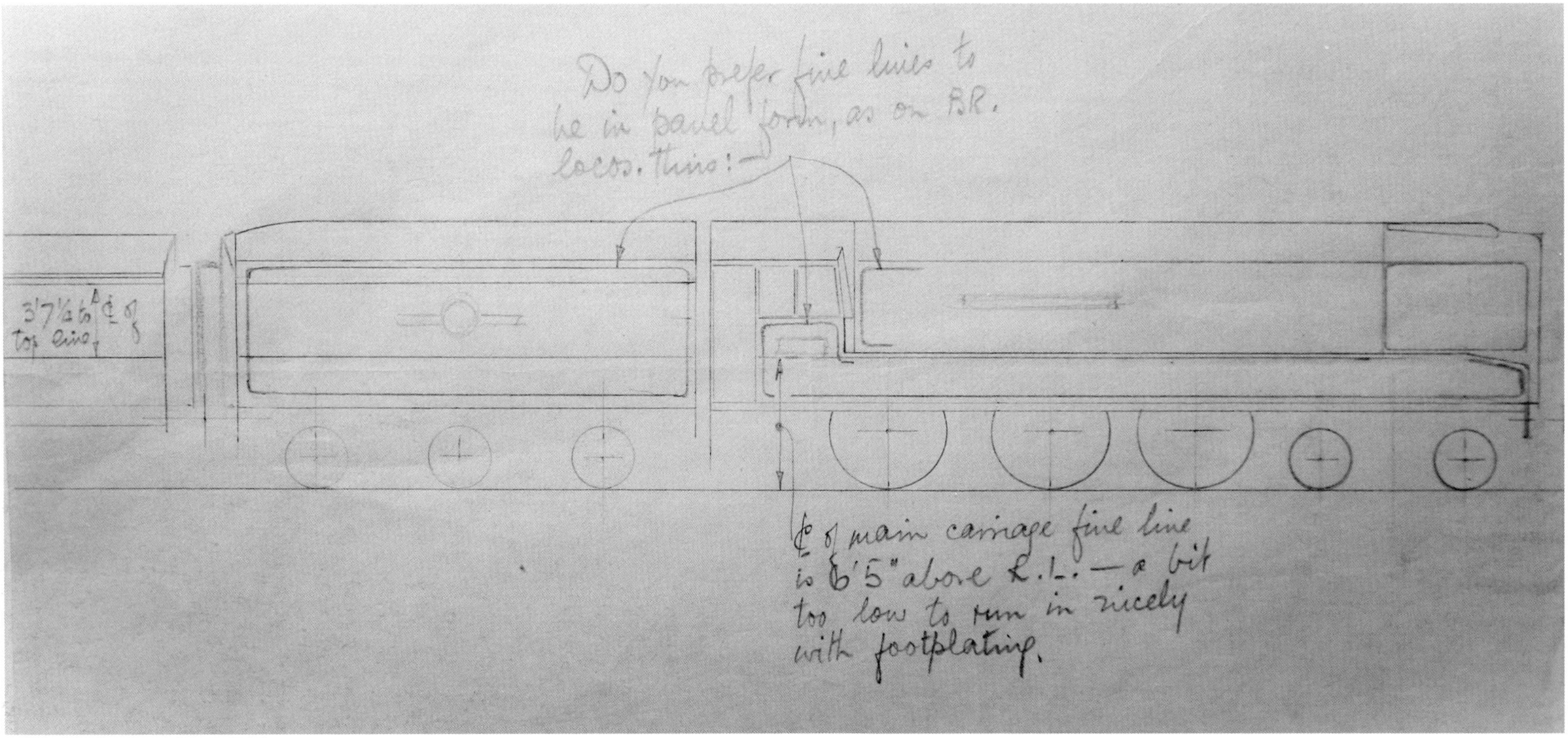

Engineering and Aesthetics – How GT3 Took Shape • 153

"This is how I visualise the front end at the moment" 1.5.56

***Opposite below, above left and above right**: **Although the GT3** project was slow to reach fruition this didn't deter Hughes and Allen from continuing to speculate on the engines looks, gradually modifying the design. In particular, they gave much thought to its front end as engineers attempted to fit the air intake filter, oil cooler, fuel and oil pumps and the alternator into a comparatively small space. These are just three of many sketches produced at this time and show the streamlined nose gradually become more 'squared off', as Allen notes in a letter to Hughes. (JH/WA)*

In the later stages of the project Allen was employed on a commission basis, receiving fees for his work. This led to him producing a series of detailed paintings to be used, one assumes, to sell Hughes's ideas to English Electric's senior managers and potential customers. It doesn't seem as though Hughes gave Allen any specific instructions on colour schemes to be used, allowing his friend to visualise the engine in a variety of paint schemes – green, brown and in the above example a red colour similar to that used on some of Stanier's Princess Royal and Coronation Pacifcs. In each case, Allen has applied the English Electric logo to the tender. (JH/WA)

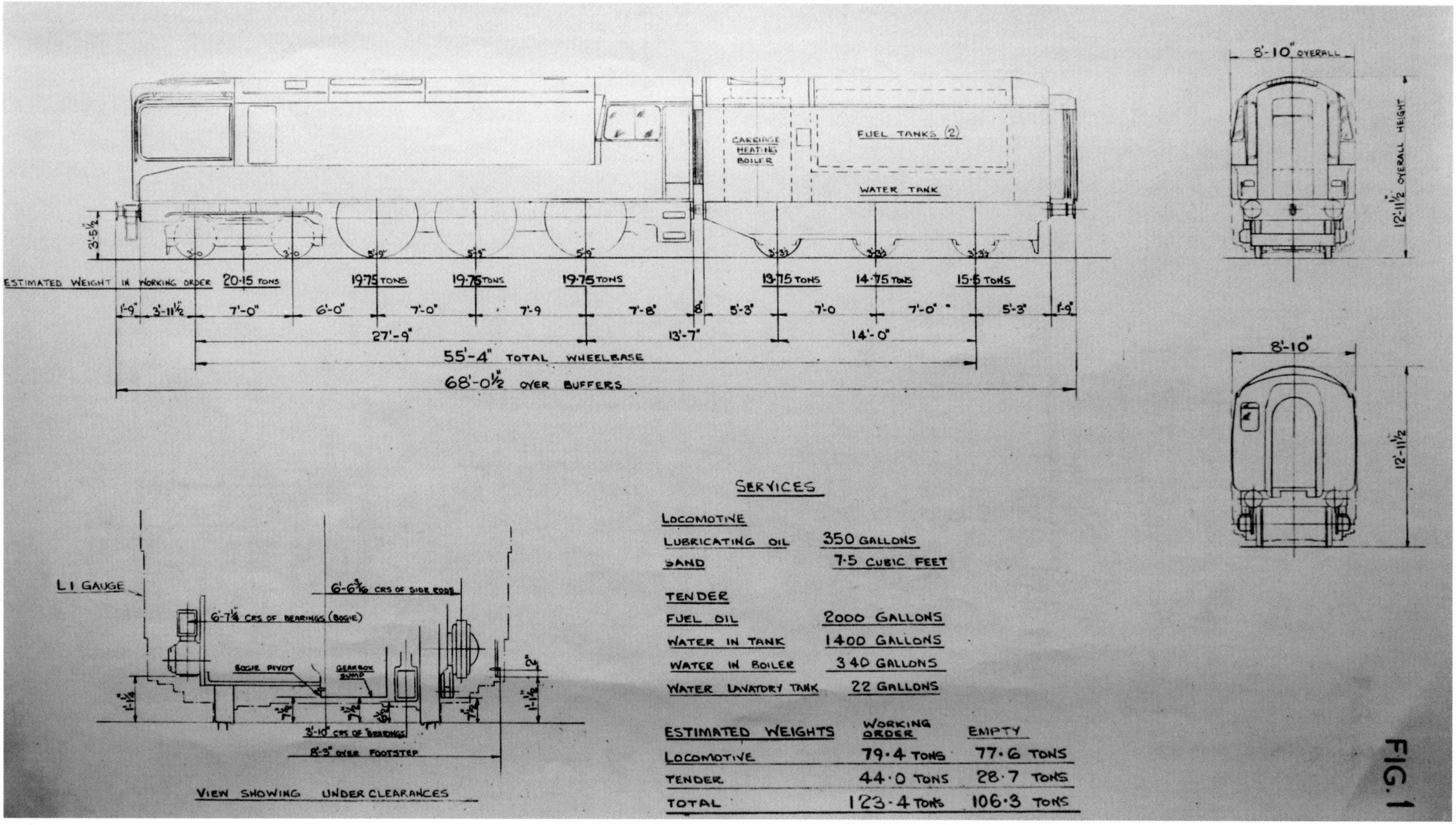

Above: **Finally, in 1955,** all Hughes and Allen's effort comes together and the final design of GT3 reaches fruition. There will be some small refinements as construction begins and then testing takes place, but in essence this drawing is the final product of many years of thought and a great deal of planning. (JH)

Right: **It isn't known** whether Bill Allen saw GT3 in action or even rode on her footplate, but John Hughes did send him this photo adding the words, 'Finally there! Though the looks are not quite as good as we hoped they would be. In time, I hope to change the cab arrangement and fit more attractive side windows – similar to those on a Merchant Navy Pacific.'. (JH)

Appendix 2
GT3 UNDER TEST AT RUGBY

Although GT3 had a very short life, the engine, most unusually, underwent a period of detailed technical evaluation at British Railway's Rugby Locomotive Testing Station as part of its development programme. In so doing, the engine was subject to a largely independent review that provides us with an interesting insight into the project's worth and any issues BR might potentially raise when considering whether to purchase this engine or when contemplating the wider application of turbines across its fleet. Parts of the draft report given to John Hughes are quoted in the main body of this book, but here, as a fitting memorial to GT3's Chief Designer and English Electric, the draft is reproduced in its entirety.

During the weeks that GT3 spent at Rugby, John Hughes was a regular visitor and collected many photos of the work underway, participating in much of it personally. On the back of this print he has written, 'Talking to Test Centre managers before the trials begin. Engine in incomplete state.'. (JH)

BR Locomotive Testing Station, Rugby
Preliminary Performance and Efficiency Test of English Electric's GT3
February 1959

Foreword

The work carried out at Rugby differed appreciably from any test previously carried out there in several respects:

- In the first place the locomotive was not the property of the British Transport Commission, but of the English Electric Company Ltd and, in consequence, that company's staff took part in the work together with the staff of the Testing Station.
- In the second place the locomotive was in an incomplete condition, the design being unfinished and subject to development in certain respects, so that a large element of the proving was involved.
- In the third place this was the first occasion on which the Rugby Test Plant had been used for trials of other than a steam locomotive.

The Locomotive (see Figs 1 and 2 for specific details about GT3)

Is of the 4-6-0 type, the six-coupled wheel being driven by the special locomotive version of the English Electric EM-27 gas turbine engine through a reversing gear and reduction gear flexibly connected to the centre axle.

The engine is of the type in which there are two turbines one connected to the air compressor and the other the useful power output. The locomotive version of the EM-27 is fitted with a heat exchanger and the power turbine has special blading to allow for its use at variable speed and to give a good starting torque when stalled. With this arrangement the first turbine and compressor, the charging set, can be kept running when the locomotive and the power turbine are stationary. The gear drive is a particular feature of the locomotive and the gear-box, which has a fixed reduction ratio, was designed specifically for the locomotive, incorporating a number of features unusual in traction applications.

The final flexible drive is of the floating ring type, which is fitted to many electric locomotives. This drive allows for vertical movement of the axleboxes in the guides, for mis-alignment of the axle and gear, due to the rolling of the locomotive on its springs and for small relative axial movements of the gear and axle. The flexible drive and gear box have a common sump.

The locomotive will eventually be equipped with a tender carrying the fuel, train heating boiler etc, but this has not yet been built. The fuel capacity will be sufficient for a journey from London to Edinburgh or Glasgow and back with a full load.

Running In

Whereas the turbine set had been run for a considerable period on a test bed at the maker's works the gear drive has not been run in, so that one of the first requirements was to run it in, with progressive increases in load and speed, whilst keeping a careful watch on oil and bearing temperatures and examining the gears at intervals.

Hughes has written on the back of this picture, 'GT3 plugged in and ready to go. A few problems had to be sorted out and mods made before work could begin in earnest. Only the most basic controls installed on the footplate for the driver to use.'. (JH)

This running in was achieved without trouble as far as the gears themselves were concerned, but a good deal of trouble was experienced with the lubrication system.

Lubrication

The lubrication system comprises a large container from which oil is drawn by the pressure pump and delivered to the turbine bearings and to the bearings of the gear box and flexible drive and to the gear teeth through suitable spray nozzles. The whole of this oil collects in the sump of the gear box from which it is evacuated by the scavenge pump, which delivers it to the top of the main tank by way of the cooler. The oil de-aerates in an enlarged upper part of the main container.

In order to avoid the use of two different varieties of lubricating oil for the main components this common lubrication system serves both the gear box and the turbine set, using oil specially blended for the purpose, this being a good deal heavier than that normally used for gears and especially for traction gears. This use of forced lubrication for the final gears and flexible drive is believed to be unique as a traction application.

There is a special 'towing pump', mechanically driven from the transmission, which functions when, for any reason, the locomotive is moved 'dead' by another locomotive. To ensure that this works, even if the main lubrication pumps are not in action, the system is designed so that as soon as the main pumps stop a suitable quantity of oil flows from the upper part of the main tank into the sump of the gearbox. This pump was removed during the tests at Rugby.

The most important of the troubles with this system were a consequence of the inability, in certain circumstances, of the scavenge pump to maintain a 'dry' sump, that is to keep the oil level at the bottom of the sump below the teeth of the final drive wheel and below the level swept by the parts of the flexible drive. As soon as appreciable splashing commenced, above a certain speed, the condition rapidly worsened until large quantities of oil were forced into the upper part of the gear box and considerable quantities escaped through the seals around the axle, which were intended principally to prevent the ingress of dirt, not to contain oil in bulk in a state of violent agitation.

After a number of attempts to cure this problem by changing the relative capacities of the pressure pumps, which retarded the onset of the trouble but did not prevent it, a cure was found by modification of the sump bottom, which brought the suction pipe of the scavenge pump in from below instead of from above. This temporary alteration of the existing bottom involved fouling the loading gauge between the rails, but this will not be necessary once a new sump is made,

Another trouble was the result of oil failing to de-aerate if the locomotive was opened up too quickly when the oil supply was cold. This caused a froth of oil to escape from the vent at the top of the main tank. This was particularly liable to occur after the locomotive had stood over the weekend in cold weather. It would be unlikely to occur with a locomotive in normal service when intervals between duties would rarely be long enough for the oil to cool down so far. The trouble was dealt with partly by care in running the locomotive after starting and by fitting a temporary heater coil in the oil tank.

A similar problem, due to low temperature, was the very high resistance to oil flow through the cooler on start up. This caused several burst flexible hoses until an adequate bypass to the cooler was installed.

One reason for mentioning these troubles is to emphasise the benefits of carrying out such development work on a test plant rather than on a locomotive in service on the line. In each case the locomotive was stopped before any damage could be done, but it might have been otherwise on the line. As it was it merely caused considerable delays to the tests.

Indeed, it is not too much to say that the overcoming of this difficulty without all the attendant troubles that might have followed had the locomotive been in service, has alone justified the policy of using a test plant for such development work, even without the other benefits.

It is quite possible that a number of developments in the past, in the course of which teething troubles occurred, were abandoned for the lack of such a facility and the consequent difficulty of diagnosing the trouble or finding a cure.

Batteries

The batteries were capable of supplying current for several starts but only for two full shut down periods of three hours, without charging, but will be able to cope with several when the demand has been greatly reduced by installing a much smaller pump for use during shut down, which tests have shown to be practicable. Normally charging would take place whenever the turbine was running but the dynamo had not been installed at the time of the tests and thus much delay was caused by stops for battery charging using a relatively small portable charging set. This difficulty was greatly reduced when a motor-generator set was provided to supply the electrically driven oil pumps, instead of using the batteries.

The Turbine

Whilst there was considerable experience of bench tests of this and similar EM-27 engines this was the first occasion of testing a complete installation and especially as it might be affected by the particular application, its response to the controls and so on.

During the whole series of tests three major troubles occurred and some minor ones.

Two of these were 'rubs', that is actual contact between the rotor and stator of the turbine, resulting in more or less severe damage to the blading. One rub occurred in the turbine of the charging set and one in the power turbine. The former was the most severe and was due to the omission of one machining process during manufacture. The casing thus became slightly over-heated and so distorted enough

Hughes wrote on this print, *'Running at high speed early in the trials. Very noisy in the chamber'*. (JH)

to take up the clearance provided. The subsequent performance of the engine was thus affected to some extent in quality and range; the full power output no longer being attainable.

The other major trouble was a consequence of faulty manufacture of one bearing housing, which caused a progressively increasing leak of lubricating oil into the stream of hot air and combustion products passing through the turbines. The leak was known to exist at the start of testing but a calculated risk was taken that it would not become serious within the duration of the tests. This proved not to be the case, at least in part because the tests were more prolonged than had been expected. Eventually appreciable quantities of oil penetrated to, and burned in a part of the casing not intended to be in direct contact with high temperature gases. As a result, several minor fires occurred. This defect was put right at the same time as the damage due to the rub in the charging set turbine.

There was also some trouble due to the accumulation of carbon on the burners in the two combustion chambers, the main effect of which was to make starting difficult or even to prevent it. Some damage was also done to the linings of the combustion chambers, but none had to be replaced.

Auxiliaries

The principal auxiliaries consist of the starter, the fuel and lubricating oil pumps, such items as vacuum brake equipment, train heating, sanding gear etc, not yet having been fitted.

All the main components worked well throughout but there was trouble due to the failure of the flexible couples between the starter and the engine, which also take the drive to the mechanically driven lubricating oil pumps, the fuel pump and the fan of the oil cooler. The trouble was traced to there being a greater misalignment in the coupling than specified. The flexible couplings were replaced by a more robust type to enable the tests to be completed and a re-design of the drive has since been made and a specimen is under trial on a test rig.

The Test Programme

The original intention was to cover as wide a range as possible of power and speed within the limits set by various considerations. But after the initial running in period it was proposed to proceed by degrees to the extremes of pull, speed and power, first separately and then together.

The upper speed limit was set by the maximum designed speed of the power turbine, which gives a maximum permitted speed of 90mph for the locomotive, which is intended to be a mixed traffic type. The lower limit of speed, at each power, would be set either by the capabilities of the test plant brakes or by the limit of adhesion as the case might be. In fact, the lowest speed, for one only, was 10mph and 15mph for several tests and similarly the highest was 90mph for one test only and 80mph for several other tests.

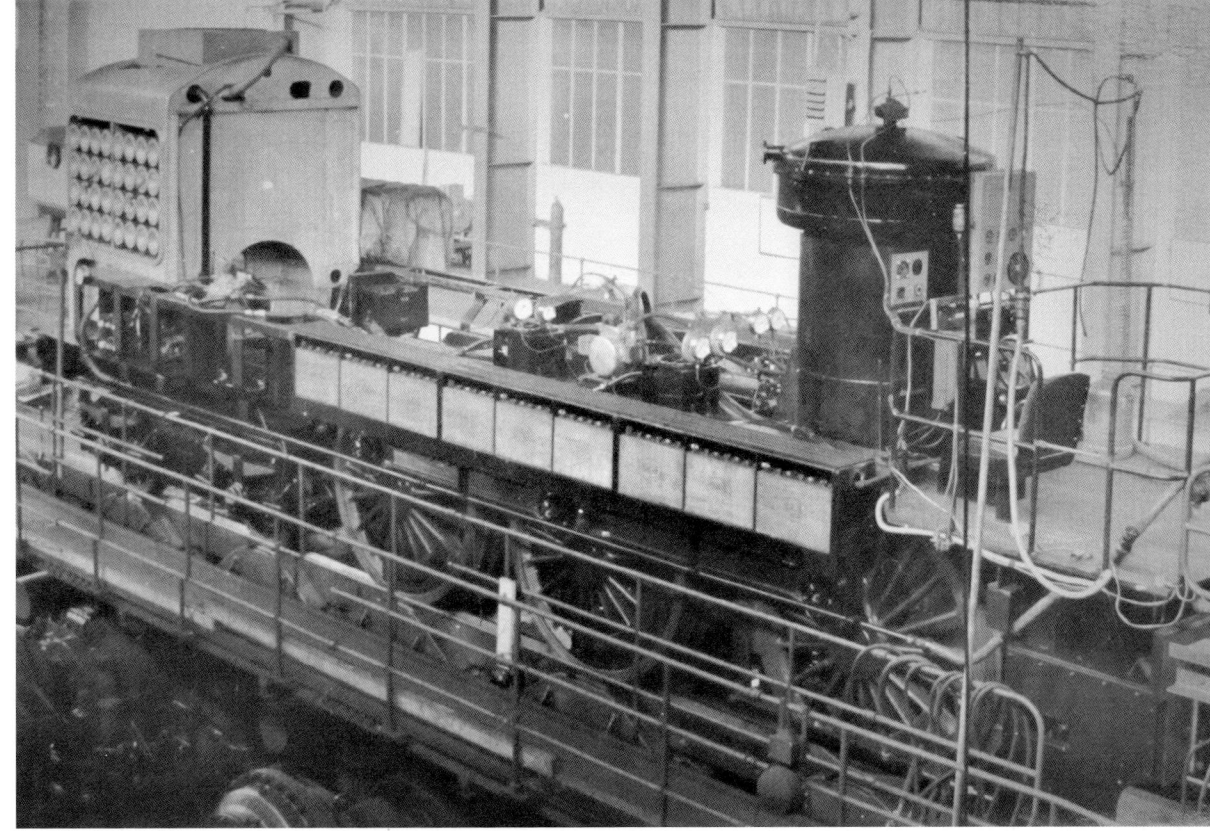

'At rest, stripped down and awaiting some attention before next trial' according to notes kept by John Hughes. (JH)

'Testing on the line (around Rugby, but not on the main line). Loco gave good account of herself pulling two of Stanier's best with their brakes on. I took the controls, briefly.' (JH)

The maximum pull was limited in the same way as the minimum speed and, as the response of the whole machine in the event of a slip occurring was unknown, it was deemed imprudent to attempt the absolute maximum at this stage.

The maximum power is limited in two ways, a maximum speed for the charging set and a maximum temperature of the gas entering the turbine of the charging set. The maximum designed speed is 8,250rpm but there is no definite figure for temperature, the actual setting being decided by consideration of the expected life of the turbine, the nature of the duty cycle of the locomotive and so on. The maximum temperature control is by regulating mechanism, which shuts down the fuel supply if the temperature is exceeded. The regulating mechanism was adjusted during the course of the tests as it originally operated well before the maximum designed charging set speed was reached.

The maximum designed charging set speed was never quite reached as the damage to the turbine of the charging set occurred before this was attempted and, after the refit, the shortening of the blade tips prevented the set from attaining its full rating by an appreciable margin.

The maximum speed of the charging set for which test figures were obtained was approximately 8,100rpm, and whilst the difference between this and 8,250rpm seems small the effect is considerable as the actual power output varies very nearly as the cube of the charging set speed. This rate was not held long enough for accurate measurement to be obtained.

After refit the maximum charging set speed was approximately 7,700rpm so that the power was very considerably cut down as to its upper limit.

Evaluation of Test Results

Essentially the evaluation followed the same methods as used for steam locomotives. The basis of the whole process is the production of a family of 'Willans Lines' each one being a plot of power against hourly fuel consumption at one speed. Such Willans Lines were obtained for speeds of 15, 20, 25, 30, 40, 50, 60, 70 and 80mph and a number of fuel rates spread over a range from 700lb/hr upwards, except at the highest speeds, at which the lower limit was rather higher. The upper limit was just over 1512lb/hr at 70mph after it.

Test Results

Before considering the actual results, shown graphically in Figs 5 to 10, it is important to define clearly just what is being portrayed. The power that is measured on the test plant is normally referred to as the wheel-rim horsepower but it is necessary to define this still further to avoid confusion.

The measured power is that emerging at the rim of the wheels after friction in the journal bearings of the driving and coupled axles has been overcome, but it does not include that component of track resistance that is due to any further depression of the track behind the leading carrying wheels, if any, as there are in this case.

To arrive at true drawbar horsepower it would be necessary to deduct not only this additional part of the tracks resistance for the driving and coupled axles, but also the whole of the track and journal resistances of the bogie and tender, respectively, together with wind resistance of the entire locomotive.

The W.R.H.P. used in the present context differs from that sometimes used in electric traction calculations because the friction in the main journal bearings has been overcome before the measurement is made.

Curves of the wheel-rim horsepower speed are shown in Figs 5 and 6. Those in Fig 5 are more accurate than those in Fig 6, which were based on partial information only and are of a preliminary character. The latter curves are included mainly because they give a good indication of the full capacity of the locomotive in its original condition and of the consequences of the shortening of the turbine blades during the refit and after the rub, which occurred as described earlier. No attempt was made during the refit to do more than clean up the damaged blade ends, which were appreciably reduced in height as a result. The final effect was to limit the maximum power output to about 80% of that originally available.

Figs 7 and 8 are the corresponding curves of pull at the wheel-rim against speed, the same remarks applying as for Figs 5 and 6. Fig 7 also shows contours of efficiency plotted in the manner made familiar in the locomotive testing Bulletins published by the BTC, but the efficiencies are not strictly comparable as the curves in the Bulletins are not relevant to efficiency at the wheel-rims, but to that at the drawbar or at the cylinders of steam locomotives.

Figs 9 and 10 show the derived values of specific fuel consumption. Again the same remarks as for Figs 5 and 6, and again these specific consumption figures are not directly comparable with figures based on drawbar horsepower.

Comments on Test Results

The shape of the curves are such as to be expected from a motive power unit of this sort.

Firstly, it is normal for a gas turbine engine to show to best advantage at full load when efficiency highest and specific fuel consumption lowest. Secondly, with a direct drive from the turbine the power and efficiency at any given flow of the working fluid will be zero both at standstill and at the runaway speed and a maximum at, or near, half the runaway speed. This runaway speed is, of course, normally well above the maximum operating speed of the machine.

As a result of these two characteristics the curves of power against speed consists of a family of roughly parabolic curves the axes of the parabolic being up and down with the vertices at the top. In the present case the curves are somewhat modified in shape because the characteristics of the transmission are superimposed on those of the engine.

It will be seen that at each fuel rate the power rises to a maximum and then falls off again as the speed rises and that the optimum speed increases somewhat as the fuel rate increases. The optimum is largely in the range of 40 to 60mph, which is very suitable for a mixed traffic locomotive. This optimum range could be either raised or lowered if a locomotive were intended mainly for express passenger or for freight service, respectively.

The maximum power output is high for a locomotive of moderate

'**Back on** the rolling road for more tests'. The figure in the white coat is thought to be John Hughes himself. (JH)

weight and dimensions, which is in no sense of lightweight construction.

The characteristics are such that the locomotive will show to best advantage if it is employed on duties involving sustained high power and mainly running between 40 and 70mph, such as heavy night trains with sleeping cars, fitted freight trains and such like. The basic concept is one of a machine for long distance working at high power.

Supplementary Test on the Line

To investigate certain aspects of operation that cannot be dealt with on the test plant some additional tests were undertaken on the tracks at the Testing Station and in the yard of Rugby Motive Power Depot and adjacent sidings. These tests had to be confined in this way as the incomplete state of the locomotive prohibited the use of running lines. For this reason, it was not thought worthwhile to make use of a dynamometer car.

Starting against a heavy load was tried by using two eight-coupled steam engines and tenders, with brakes applied in varying degrees, as the load and then starting on progressively more severe gradients. The most severe condition attempted was on a gradient of 1 in 44 combined with a curvature of 5½ chains on roughly laid track, when a start was made without slipping although the locomotive was, in its unfinished state, nearly four tons light in adhesive weight. The start was rather slow as there was an appreciable delay whilst the charging set speed was increased slowly until the torque on the power turbine rose to a high enough value smoothly, gradually increasing speed. The opinion expressed that no steam locomotive usually using this line (Class 5MT and Classes 7F and 8F) would have been able to start the load at that point on the line is pertinent. The rails were dry but not sanded.

It is a feature of the locomotive that maximum pull can be exerted at low speed without any time limit.

On two occasions when pulling the two steam locomotives on this severely graded and sharply curved track, at a place where the track was very roughly laid and at a turn out, the wheels all slipped on dry rails and on each occasion the slip stopped before it became really severe, this being achieved, if not as promptly as with steam locomotives, at least more so than with some other types.

Examination After Completion of the Tests

All the principal parts of the engine and transmission were dismantled at the conclusion of the tests, which had involved just over 5000 miles of running under power on the test plant. Prior to this the turbine had run for a considerable number of hours during bench tests and the transmission had run some forty miles or so with the locomotive in tow and subsequently a few miles were covered under power during the supplementary tests on the line.

In the summary that follows mention is made of all the features noted, including those due to the incidents mentioned above, unless the context makes it clear that this is not so.

Summary of the Condition of the Engine etc at the End of the Test

Engine

Compressor – A slight rub had occurred between the impeller vanes and the diffuser casing – no damage was caused. The aluminium diffuser guide vanes have some pitted areas. Thrust and journal bearings – good condition and free from scores. The rotor and impeller are free from corrosion. There was no trace of an oil leak from the front gland.

Charging Turbine – No further blade rubs have occurred. Modified intermediate bearing block free from oil leaks and the glands were clean. Turbine inlet liner requires clearance for axial expansion, the inlet make up pieces are distorted and have pressed heavily on the liner. There is some distortion in the turbine inlet casing but none in the outer ring or stator rings.

Power Turbine – Light rub between stator blades and rotor – no pick up. Light rub between first stage rotor blades and stator packers – no pick up, also on second stage with several blades slightly bent. The third stage is free of rubs but one has bent blades. The third stage shroud ring is crushed and distorted and requires re-design.

Pressure Sump – Float valve tappet screw badly worn.

Combustion Chambers – Burnt patches on louvre wall, carbon build up on burners

Heat Exchanger – Crack in each of the bottom corners which will be modified to a later design. Crack in bottom side cowl due to contact with exhaust chamber.

Transmission Gearbox

Gearing – The gearing generally appears to be satisfactory with good bedding, the primary helical gears are in good condition with the exception of the marks on the reverse pair due to the previous roller bearing cage failure (before the locomotive reached Rugby). Since the marks are on the following flanks they can be regarded as satisfactory. The spiral bevel gears are in excellent condition. The final spur gear wheel shows slight wear; this material is, however, of a high tensile strength and is meshing with a hardened and ground pinion, the slight wear therefore is probably confined to the initial running.

Bearings – The high speed shaft thrust and journal bearings are in excellent condition. All the roller bearings associated with the spiral bevel gears are satisfactory and there is no repetition of the overheating of the bearing cage within the helical wheel since the modification to the oil feed. The quill gear bearings are satisfactory and there is evidence that the split in the bearing shell can be positioned in the normal relation to the cap. The modified oil supply to the high speed quill shaft actuation bearing is successful and the bearing is in good condition.

Final Drive – The four bearing bushes of the floating ring are in excellent condition with no sign of wear. Due to a failure to plug the hollow mounting bolts, lubricating oil had reached the outer surfaces of the rubber spring unit with consequent deterioration.

Gearbox – The special high temperature oil resisting paint has peeled in places where applied to a cast iron surface – no such peeling has occurred on steel surfaces.

Auxiliaries – The running time on the auxiliaries was of course negligible compared with the proof testing to which they had been subjected to and they are as new. The cracking of the white metal in the rear pinion bearing persisted and is related to the imbalance of the propeller shaft which was not a satisfactory type.

Appendices to Report

(Figs 1, 2, 3 and 4 not included in report)

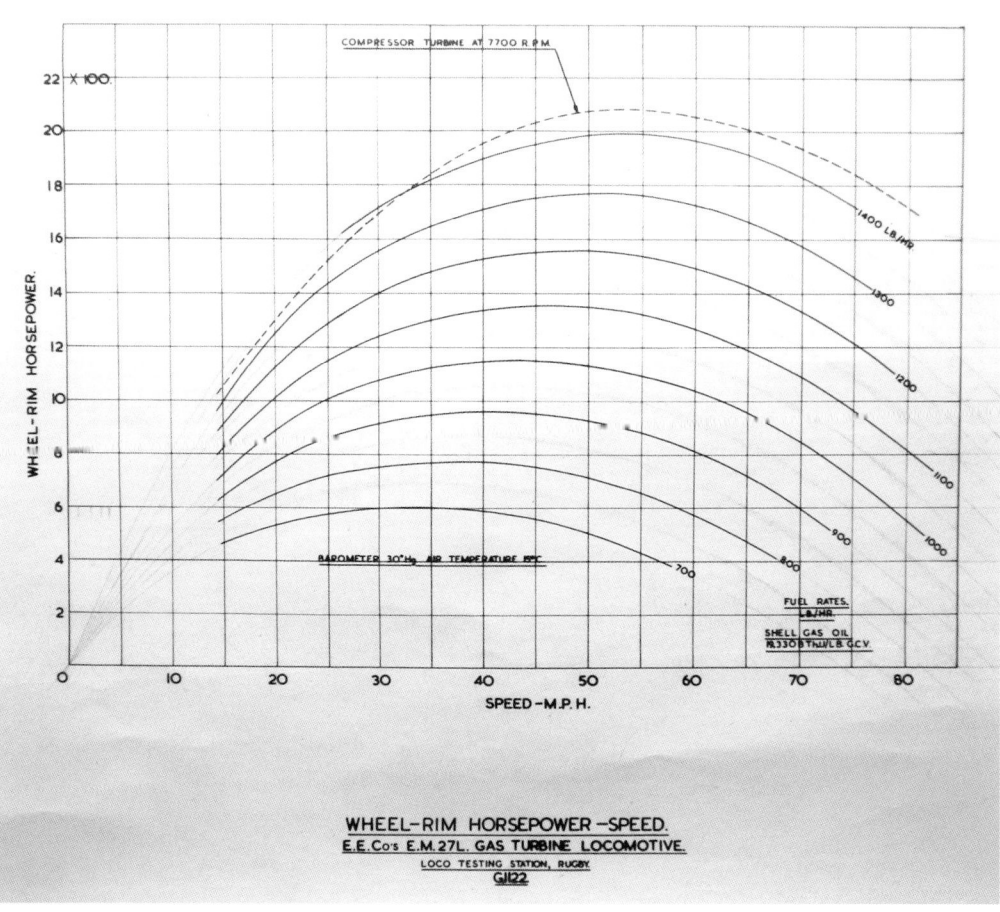

Fig 5

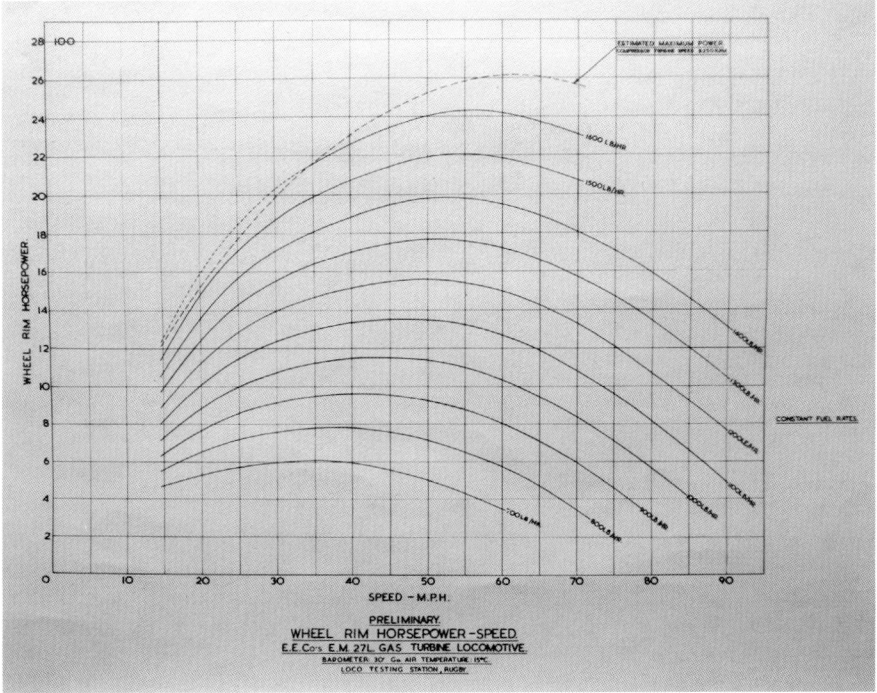

Fig 6

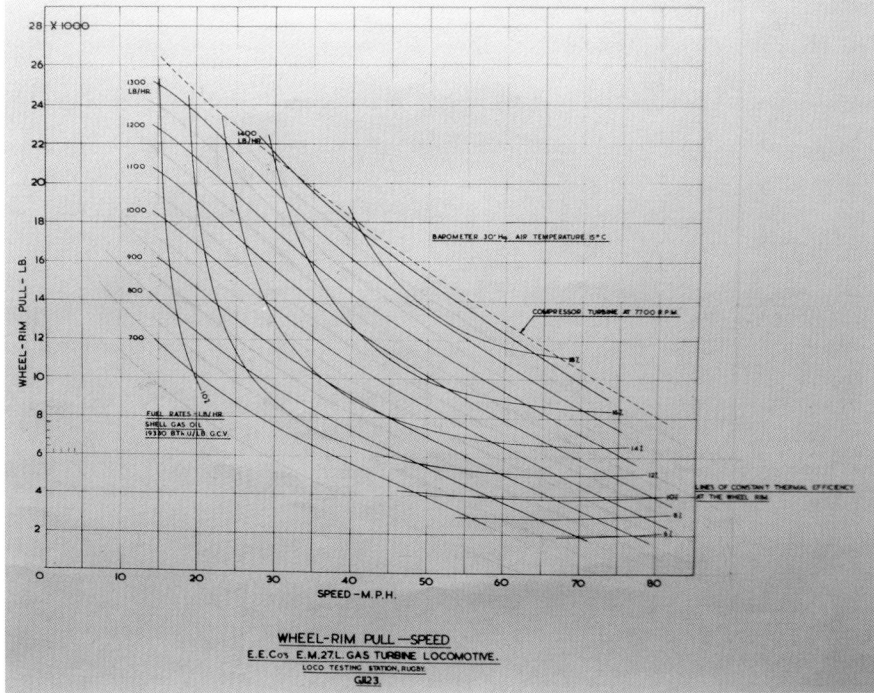

Fig 7

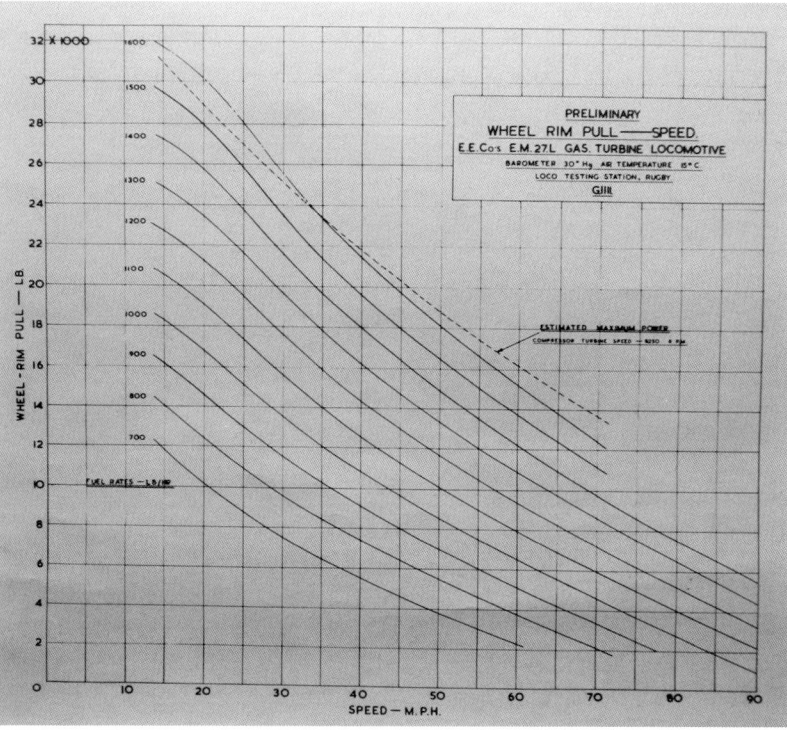

Fig 8

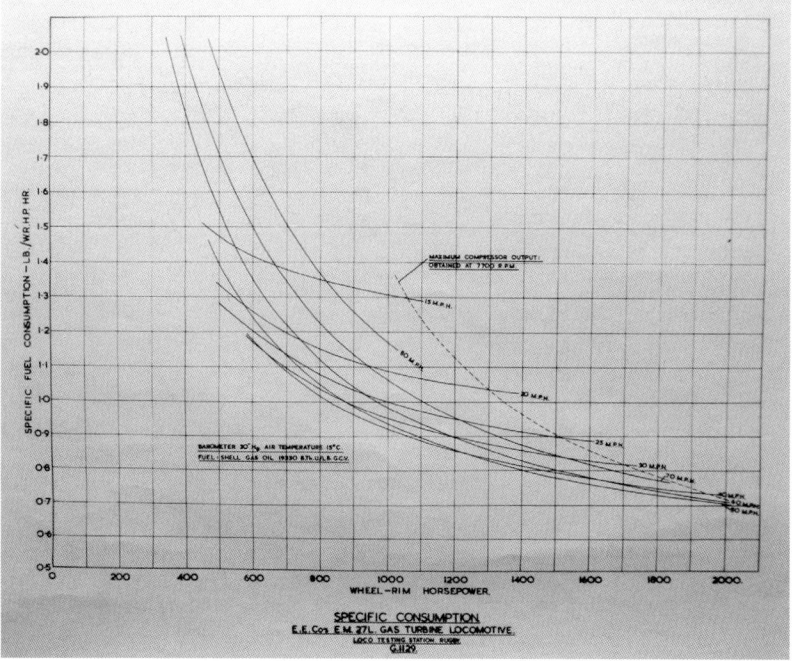

Fig 9

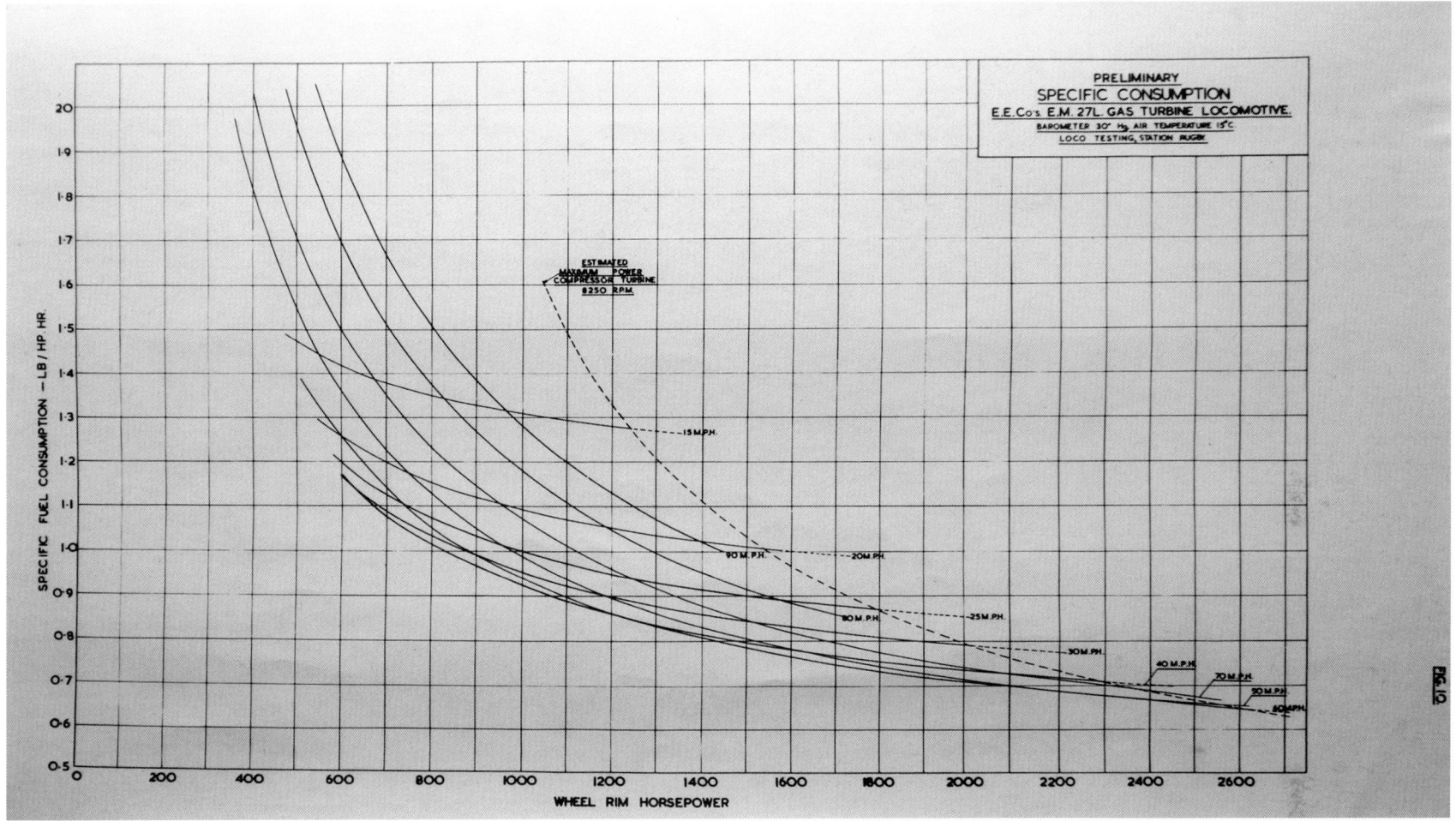

Fig 10

Appendix 3
GT3 Dynamometer Car Tests

Once GT3, in an unfinished state, had passed through the hands of staff at the Rugby Testing Station, it was hoped that progress towards completing the locomotive would be rapid. Without this she could not begin live trials on the main line with, it was hoped, a dynamometer car attached to record performance in fine detail. For one reason or another, which I have dealt with in the main body of this book, this didn't happen until late 1961. By the time the report of these trials appeared in 1962 the future of the entire project was under review. As all the costs of development and construction had been met by English Electric, the company needed an order for one or more engines to make continuing the project worthwhile. With BR wavering over whether to buy or not, much, according to John Hughes, 'rested on completing these trials successfully, or so it seemed to me'. For this reason, and considering its potential impact on the project, the report is produced here in full.

Report No: L.163
CM&EE(LMR) Derby
February 1962

Dynamometer Car Tests with English Electric Gas Turbine Locomotive No. GT3

Introduction
Following a period of running in between Carlisle and Crewe

GT3 near Shap appears to be making light work of a heavy load, although on this occasion without a dynamometer car in tow. On 14 December 1961, when undergoing test, it was a different matter. Here when pulling a fifteen carriage train, which included the LMR's dynamometer car, the engine failed to impress due to a 'deficiency in power output'. (JH)

with trains of empty stock, the English Electric Company made a proposal for dynamometer car tests on runs scheduled for the 13th and 14th December 1961. The runs were intended to conform to the existing special timings, which involved fairly high average speeds. No special instruction was required, as it was not intended to do more at this stage than obtain dynamometer car records of speed, drawbar tractive effort and horse power, with 15 coaches (including the dynamometer car). It was anticipated that more detailed performance tests with the Mobile Test Plant would be considered in 1962.

Test on 13th and 14th December 1961

On arrival at the Diesel Traction Depot, Crewe, at 7.30am on the 13th, it was found that the batteries were flat and the turbine could not be started. The test run was cancelled for the day whilst the batteries were re-charged and the reason for the low state of charge investigated.

On the 14th, the scheduled test run was made, with a load of 466 tons, including No.3 Dynamometer Car. The departure from Crewe was 15 minutes late.

When under way, the speed and power were progressively increased, but it was immediately apparent that at full controller the compressor speed was 400 to 500rpm lower than the designed speed of 8,250rpm, and the power output was accordingly considerably lower than the designed value. The discrepancy was attributed by the English Electric technical staff to premature operation of the temperature trips designed to increase the spill return from the fuel nozzles, and there by limit the combustion fuel whenever the gas temperature at the turbine inlet approaches the maximum safe value of 777 degrees C.

A fair amount of running at full controller was obtained, but this was mainly at speeds above 50mph, and in all cases was subject to a lower power as a result of the deficiency in compressor speed. The reduced power, as well as a number of unforeseen stops, etc, resulted in a very late running (over one hour). Whilst it was intended, therefore, to reduce the speed to about 10mph and obtain drawbar measurements at full controller at the steep rising gradient from Tebay onwards, this was not attempted, to avoid further loss of time, and the ascent to Shap Summit was made as fast as possible. The speed balanced at about 20mph, instead of 38mph, which was the estimated balancing speed for this weight of train on a 1 in 75 rising gradient.

It was intended to utilise the falling gradients beyond Shap Summit for coasting tests with the power handle in the 'off' position, in order to obtain data for the estimation of locomotive resistance. This, however, was impracticable, since the drive cannot be disengaged while running. It was, therefore, decided to reduce the compressor speed to 4,000rpm and to test it in this condition. However, the driver inadvertently allowed the compressor speed to drop to 3,300rpm, and in attempting to correct this caused the compressor to surge and fall to 1,500rpm, the turbine stop button was pressed, and the train was brought to rest to allow the turbine to be restarted. The attempt to carry out a coasting test was, therefore, abandoned.

At the start of the return run from Carlisle, a small fire occurred in the driving cab, which was responsible for another late start (one hour). The cause of the fire was overheating of a starting resistor for a lubricating oil pump motor. This should have cut out once the motor was up to speed, but had failed to do so. In order to avoid a recurrence of the trouble on the return run, care was taken to prevent the electrically driven auxiliaries starting up. This normally occurs when the compressor speed falls below 5,000 to 6,000rpm replacing the engine driven auxiliaries which operate at higher speeds. To maintain the compressor speed above 6,000rpm, the controller was left in the fully open position for extended periods, control of the locomotive being carried out almost entirely on the brakes.

It was found to be impossible to start the train of 15 coaches on the 1 in 131 rising gradient at Carlisle (Upperby Bridge Junction), and the assistance of the banking engine was required at the rear. The opportunity was obtained on this occasion, however, to record the starting tractive effort of the locomotive at various compressor speeds. On the 1 in 131 gradient, this ranged from 12,200lbs at 7,000 rpm to a maximum of 20,900lbs at 8,250rpm. These values become 14,000lbs and 23,100lbs respectively when corrected to level conditions. At Lancaster and at Preston some low speed tractive efforts were recorded with full controller setting, but such measurements were accompanied by rapid acceleration as the track was fairly level.

After a poor performance on 14 December 1961 the manager in charge of the tests reported that, 'it is considered that further dynamometer car tests would not be justified until the control system has been sufficiently developed to produce a reliable output'.
It was a criticism that continued to echo as BR contemplated whether to purchase this engine or not. This photo taken at about this time as GT3 rests between turns. (JH)

Results
It is normal to present information on the tractive capabilities of a locomotive in the form of drawbar tractive effort – speed curves, the tractive efforts being applicable in constant speed conditions on level track. The actual drawbar tractive effort which is measured on the dynamometer car is the effort existing at the drawhook under the prevailing conditions of gradient and acceleration, and requires correction in these two aspects to produce 'tractive effort at constant speed on the level'.

In view of the somewhat variable performance of this locomotive this procedure has been omitted, and the results are shown in Fig 1 of this report as 'actual' tractive efforts. So that the deficiency in power can be readily appreciated, a curve of 'actual' tractive effort has been calculated for the train load used on the test, based on the makers' designed full power characteristic, together with an estimated locomotive resistance. This curve is also shown in the graph.

A continuous log of controller setting, compressor speed, fuel, burner and lubricating oil pressures, and various engine and gearbox temperatures was made throughout the run by an observer on the locomotive. These figures have not been included in this report, but can be made available if required.

Conclusions
With the locomotive in its present condition, the power which is obtained at the drawbar with the driver's controller fully open, is subject to wide variations due to variable compressor speeds (7,500 to 8,250rpm). The variation

increases with speed, and amounts to about 400hp at 50mph.

At certain times the compressor speed rose to its full value of 8,250 rpm, but even on these occasions there was a serious deficiency in power output amounting to 330hp and 50mph.

In view of these undesirable features, it is considered that further dynamometer car tests would not be justified until the control system has been sufficiently developed to produce a reliable output.

Fig 1 (Report L163)

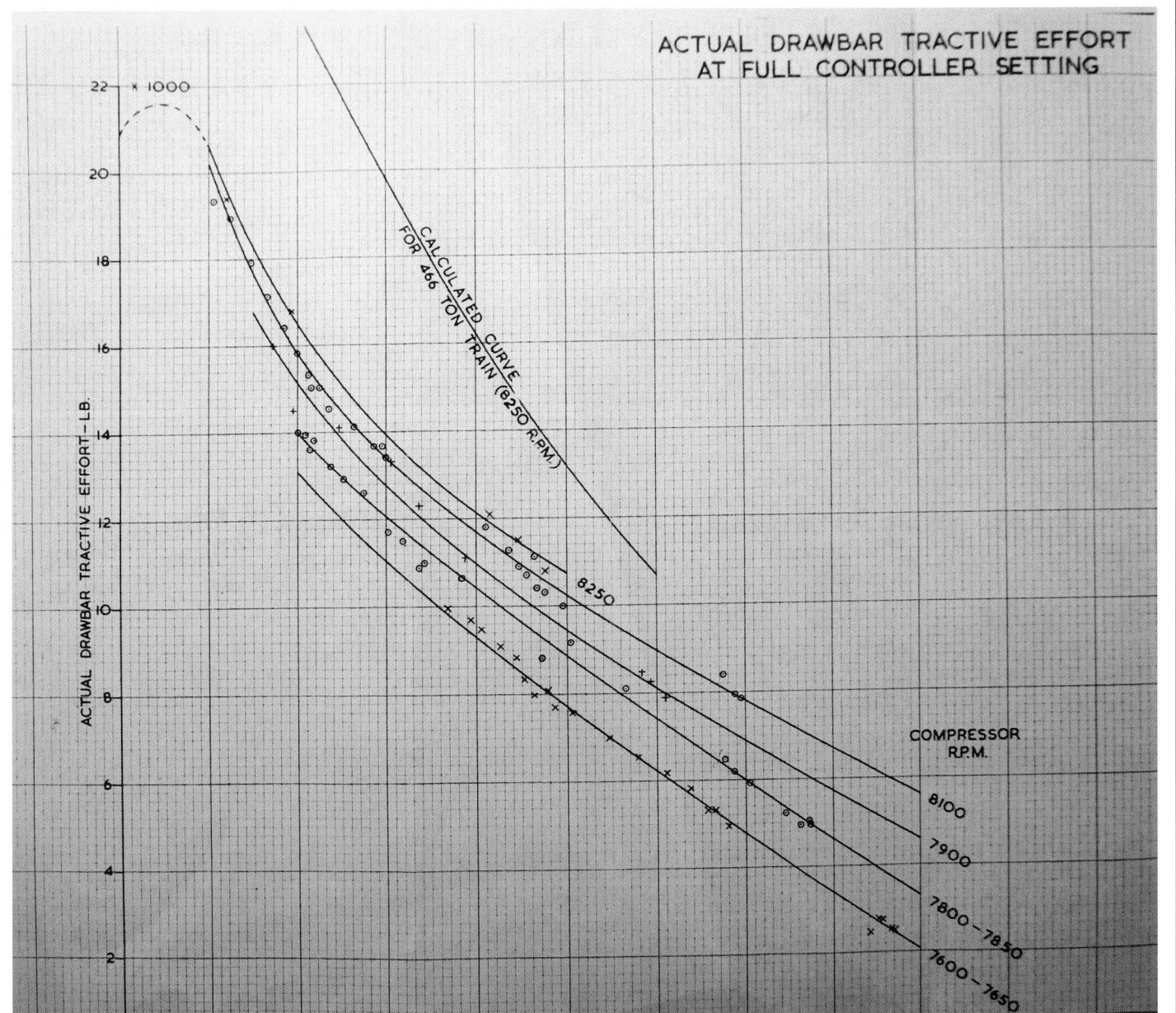

Appendix 4

JOHN HUGHES'S PROJECT SUMMARY AS PREPARED IN APRIL 1962

GT3 in its final form, according to John Hughes, in late 1961 when hope still existed that an order from BR might be forthcoming. Throughout the year Hughes avidly collected data he hoped might persuade a reticent customer to make a leap of faith. This was not to be and Hughes then worked his statistical analysis into an 'obituary' for the project instead. (JH)

Engine Data at 31st December 1961

Type	4-6-0.
Weights:	
Locomotive	79½ tons.
Tender (including 2,000 gallons of Fuel and 1,700 gallons of water)	44 tons.
Adhesive weight	59 ¼ tons.
Factor of adhesion	3.7.
Maximum starting tractive effort	36,000 lbs.
Overall length, over buffers	68ft 6½in.
Maximum Width	8ft 10in.
Maximum height	13ft 0in.
Driving wheel diameter (new)	5ft 9in.
Bogie wheel diameter (new)	3ft 0in.
Tender wheel diameter	3ft 3 1/2 in.
Maximum curve negotiable	4½ chains.
Maximum speed	90mph.
Brakes	Vacuum.
Battery (54 cells)	.550 amp hours.
Lighting and starting circuit	110 volts.
Carriage heating boiler	Vertical tube, manually Controlled 2,000 lb/hr

Power Turbine Data

Type – EM-27L open cycle two-shaft recuperative gas turbine engine:

Power turbine output	2,700hp.
Maximum speed of power turbine	9,000rpm
Maximum speed of charging turbine	8,250rpm.
Maximum mass flow	42.5 lb per sec.
Design turbine inlet temperature	777 degrees C.
Exhaust temperature at full power	355 degrees C.
Idling fuel flow	0.6 gpm.
Specific fuel consumption at full power	0.635 lb/bhp/hr.

Summary of Test Running up to 31st December 1961

1. At British Railways Locomotive Testing Station, Rugby

Distance covered	Engine Running Time	Loads	Remarks
5120 miles – max speed 92mph.	159 hours	Nil	Dynamometer loading and short track runs.

2. Between Whitchurch and Llandudno (70 miles each way).

Distance covered	Engine Running Time	Loads	Remarks
1722 miles – max speed 85mph.	100 hours.	8 to 16 coaches (255 to 510 tons).	Branch line running & driver training Included.

3. **Between Leicester and London (103 miles each way).**

5508 miles – track speed limit – 75mph.	230 hours.	6 to 9 coaches (190 to 285 tons).	This includes 3 weeks of runs to Woodford Halse only, for driver training. Timings were to normal express standards.

4. **Between Crewe and Carlisle (145 miles each way).**

3520 miles – max speed 85mph.	140 hours.	10 to 15 coaches (320 to 480 tons).	Times were to normal express and fast express standards.

The above table only shows the actual running of a completed locomotive. Separately from this, the engine was run in its prototype and final locomotive versions as follows:

Prototype Engine EM-27A – 1800 hours on dynamometer loading, including 1000 hours at full load and cycling tests (660 cycles) between 6000 and 8250 rpm.

Prototype Engine EM-27L - 237 hours on bench tests for mechanical proof and performance.

The company also has experience of this type of engine with its industrial version, EM-27P.

Of these three versions the one most used, being on non-stop duty day and night, has been installed at the Fawley refinery of Messrs Esso Ltd. At the date of writing it has completed more than 24,800 hours of work since installed in November 1958, with availability running at 99.7%. This engine is operating under ideal conditions, ie with a fixed installation, no heat exchanger and at from 75% to more than a 100% load. The maintenance costs have been negligible.

Table of Defects Arising in the First 11,000 miles of Operation

Reversing System	Modification
1. The lower arm of the reversing gear in the cab turned on its shaft due to bruising of over-loaded key.	New shaft fitted with serrated pinch bolted lever fixings.
2. Slips occurred between power turbine brake and rollers.	Larger capacity hydraulic brake cylinder fitted. Timing of brake application changed.
3. Balance pistons did not always rise when a load was applied.	Air valve added to bring piston to approximate position of balance before applying the brake.
4. Stop faces of forward gear pilot ring bruised causing it to remain at wrong end of stroke.	Rotation of pilot ring changed to be the same as turbine.
5. Overheating of thrust bearing between the reversing shaft and splined drive shaft.	New bearing of larger area and with more liberal oil supply fitted. Bearing Is freed of engagement load.
Fuel System	
6. Fuel leaked into sealed capsule of the spill line reducing valve, increasing spill flow.	New capsule with hand soldered joints constructed. Gauge fitted to engine fuel panel to show if change occurs.
7. Fracture in braze of nipple on burner line supply pipe due to poor penetration.	Stainless steel pipes replaced by Aeroquip PTFE flexible hose.

8. Fuel in spill line running back into combustion chamber on shut door.

   Non-return valve fitted in spill line.

9. Gasket on fuel supply pressure joint blew.

   Correct high tensile studs fitted and more positive locking for nuts.

10. Swollen synthetic rubber joint washer allowed leakage past inlet filter element.

    Replace by felt washer.

Electrical

11. Horn switch contacts burnt out.

    Air suppressors fitted.

12. Sequence switch motor some-times did not start on pressing start button.

    Additional contact fitted to start button to give momentary 24v supply to the motor.

Lubricating System

13. Bearings of towing pump damaged through overheating.

    The end float of the gears was below the design value.

14. Towing pump relief valve spindle fractured.

    This was a fatigue failure. Design and material improved.

15. The motor driven pressure and scavenge pumps did not always prime with engine running.

    Small feed provided to pump inlets from engine pressure pump. De-aerator pipes fitted to pumps.

16. Main lubrication oil tank vent within superstructure deposited oil on roof.

    Extra trap fitted and discharge of vent piped through the roof.

17. Right-hand gear case oil seal leaked with locomotive on left-hand curve.

    Stronger return springs fitted to seal and baffle fitted to towing pump relief valve discharge.

Miscellaneous

18. Air supply pipe from engine to balance cylinders fractured.

    Replaced by temperature flexible hose.

19. Ejector admission valve hunted on change over.

    Damper fitted.

20. Cracks in light gauge turbine exhaust chamber.

    Repaired with a more flexible section.

REFERENCES AND SOURCES

The National Railway Museum (Search Engine)
Records Consulted
The R. Bond Collection.
The E.S Cox Collection.
The R. Riddles Collection (donated by author).
Rugby Test Centre Papers.

The National Archives (Discovery)
Records Consulted
AN 172/261.

Other Collections Consulted
Institution of Mechanical Engineers.
Institution of Locomotive Engineers.
R.A. Hillier.
J. Constantine.
D. Neal.
T.F. Coleman/M Lemon.
John Hughes Collection (some of which is held by the National Railway Museum).

Periodicals and Journals
IMechE/ILocoE Journals
The Engineer
The Gazette various dates.
The Mecanno Magazine
Steam World
Swindon Engineering Society
The Stephenson Society Journal

Books
Allen, J.R. and Bursley, J.A., *Heat Engines; Steam, Gas, Steam Turbines and Their Auxiliaries* (1941).
Bond, R., *A Lifetime With Locomotives* (1975).
Brown, E.A.S., *Nigel Gresley. Locomotive Engineer"* Littlehampton Book Services (1961)
Chacksfield, J.E., *Sir William Stanier* (2001).
Dalby, W.E., *The Balancing of Engines* (1920).
Dalby, W.E., *British Railways: Some Facts and A Few Problems* (1910).
Hillier-Graves, T., *Gresley and His Locomotives*, Pen and Sword Transport (2019).
Hillier-Graves, T., *Gresley's Silver Link*, Pen and Sword Transport (2022).
Hillier-Graves, T., *Peppercorn. His Life and Locomotives*, Pen and Sword Transport (2021).
Hillier-Graves, T., *Tom Coleman. His Life and Work*, Pen and Sword Transport (2019).
Hillier-Graves, T., *Turbomotive – Stanier's Advanced Pacific* Pen and Sword Transport (2017).
Holcroft, H., *Locomotive Adventure Vols 1 and 2*, Ian Allan (1962).
Nock, O.S., *William Stanier,* (1964)
Pope, A., *Wind Tunnel Testing* (1947).
Rogers, H.C.B., *Transition from Steam*, Allen and Unwin (1980).
Tuffnel, R., *Prototype Locomotives*, (1985).

PHOTOGRAPHIC SOURCES/CREDITS

John Hughes (JH), R. Hillier (RH), Author, D. Neal (DN) and J. Constantine (JC).

Copyright is a complex issue and often difficult to establish, especially when a photograph or document exists in a number of public and private collections. Strenuous efforts have been made to ensure each item is correctly attributed, but no process is flawless, especially when many of these items are more than 70 years old with photographers or authors long gone. If an error has been made, it was unintentional. If any reader wishes to affirm copyright, please contact the publishers and an acknowledgement will be included in any future edition of this book, should a claim be proven. We apologise in advance for any mistakes.

A number of documents held by the NRM and The National Archives have been quoted in this book. My thanks to both institutions for permission to use them.

INDEX

Advanced Passenger Train (E) – 142-147.
Allen, W (Bill) – 89-92, 99, 145, 149-154.
Allis-Chalmers Manufacturing Co – 48.
Arab-Israeli War – 143.
Argentine State Railway – 36.
Armstrong-Whitworth – 35.
Associated Electrical Industries (AEI) – 138.

Bailey, Richard – 40.
Barber, John – 43, 44.
Battersea Power Station – 32.
Bebbington, Driver – 126, 133.
Belluzzo, Guiseppe – 33, 34, 144.
Beyer-Peacock – 36, 37.
Bond, Roland – 10, 137-139.
Boveri, Walter – 47.
Boxpok wheels – 152.
BR Modernisation Plan 1955 – 29, 58, 84-86, 97, 119.
BR Nationalisation 1948 – 16.
Branca, Giovanni – 29.
Bristolian Service – 58.
British Empire Exhibition – 35.
British Leyland – 14, 143.
British Transport Commission – 93, 97, 118, 119, 122, 123, 137.
Brown-Boveri – 24, 45-48, 53, 92, 103, 123, 141.
Brown, Charles – 47.
Brush Co – 24, 83.
Brush, Charles – 30.
Bulleid O V S – 10, 146, 150, 152.

Carling, Dennis – 110, 139, 140.
Churchward, George – 23, 52.
Coleman, Tom – 11, 27.
Collett, Charles – 52.
Cook, Kenneth – 21, 22, 54.
Cox, Ernest – 10.

Dalby, William – 21.
De Laval, Gustaf – 30, 31.

Edinburgh, Duke of – 10, 11, 125, 126.
EM 27 – 27, 28, 63, 82, 83, 88, 104, 146, 158, 174.
English Electric – 14-19, 26, 44, 59, 64, 81-84, 89-99, 101, 104, 106, 109, 110, 117, 118, 121-125, 129, 131, 132, 136, 138, 140, 141, 146-149, 153, 168.
Ernesto Breda Co – 34.

Farnborough, RAE – 24, 142.
Festival of Britain – 25.

General Electric – 48, 49.
Grangesborg-Oxelosund Railway – 38, 39.
Great Western Railway – 52.
Gresley, Nigel (Sir) – 22, 24, 91, 112, 128, 142, 146.
Guy, Henry (Sir) – 39-41.

Handley, J G – 36.
Harrison, John – 131, 138.
Hawksworth, Frederick - 53, 54.
Hughes, George – 109.
Hughes, John – 11-15, 25, 26-28, 42-45, 47, 49, 56, 58, 59, 62, 64, 81-84, 87-92, 95, 97-99, 101-103, 106-108, 110-115, 118, 119, 128, 131-138, 140, 144-150, 152-154, 172-175.

Institution of Locomotive Engineers – 8-15, 23, 36, 44, 92, 97, 114, 124, 138, 139.
Institution of Mechanical Engineers – 9.
Italian State Railway – 34.
Ivatt, George – 83, 84.

Jarvis, R J – 140.
Johnson, H C – 131.
Jones, Sydney (Dr) – 142, 143.

Ljungstroms – 36-40.
Ljungstrom, Frederick – 37-42.

Locomotives:
  Electric Classes – LNER EM1/EM2 – 94, 98.
  SR Class 71 – 127.
  Diesel Classes – Brush Bo Bo (Hawk) – 24.
    Brush Type 2 (AIA-AIA) – 21.
    Class 42 Warship (Bo Bo) – 9.
    English Electric DP1/Deltic – 19, 85, 98, 101, 104, 123, 129.
    English Electric Type 1 (Bo Bo) – 20.
    English Electric Type 4 (Co Co) – 20.
    General Electric UP 57 – 51.
    LMS 10000/10001 – 83, 84.
    Union Pacific M-10002 – 61.

Index

Turbine Classes –
- Belluzzo 0-4-0 – 33.
- Belluzzo/Breda 2-8-2 – 34.
- Beyer-Peacock – 36, 37.
- Brown-Boveri Am 4/6 – 45-49.
- Brown-Boveri (No. 18000 – GT1) – 53-55, 59, 98.
- English Electric GT3 – 6-15, 23, 28, 42, 44, 87, 90, 94-100, 103-111, 112-121 (at the Rugby Test Centre), 122-136 (mainline testing), 137-141, 143-145, 150, 154, 155-167 (Rugby Test Report), 168-171 (Dynamometer Car Reports), 172-175 (John Hughes Project Report).
- General Electric UP 50 – 48, 50, 52.
- Henschel, Krupp-Zoelly & Maffei – 37, 38.
- Ljungstrom 2-8-0 – 38-40.
- Metrovick (No. 18100 – GT 2) – 56-59, 98.
- NBL Pulverised Coal Locomotive – 60-63.
- NBR Challenger – 35.
- Reid-Ramsey – 34-36.
- Turbomotive – 11, 12, 27, 29, 39, 40-42, 53, 56, 64, 93.

Steam Classes –
- BR Class 7P – 18.
- BR Class 8P – 10.
- BR Class 9F – 10.
- GWR Castle Class – 94.
- GWR King Class – 54, 94.
- LMS Class 5 – 18, 127.
- LMS 6P/7P – 127, 135.
- LMS Class 8F – 11, 117, 160,
- LMS Princess Coronation Class – 11, 12, 102, 129, 153.
- LMS Princess Royal Class – 40, 153.
- LNER A4 Class – 10, 15, 21, 91, 98, 102, 125, 142.
- LNER W1 – 128.
- SR Merchant Navy Class - 148, 154.
- Union Pacific/ALCO 4-8-8-4 'Big Boy' - 50.

M A McEvoy of Derby – 26.
Machine Tools Exhibition – 21.
Macleod, James – 35.
Marples, Ernest MP – 9.
Marylebone Exhibition – 8-15, 124-126.
Metropolitan-Vickers – 24, 32, 40, 54, 56, 57, 58, 59, 92, 93, 123, 141.
Milne, James (Sir) – 54.
Ministry of Fuel and Power – 61, 63.
Motor Rail Ltd – 138.
Mussolini, Benito – 33.

National Physical Laboratory (Teddington) – 24.
Nelson, George (Sir) – 82, 83, 88, 92, 93, 96, 97, 147.
Nelson, Henry (Sir) – 82.
Nock, O S – 132-136.
North British Locomotive Co – 34, 60-63, 83, 98.
Nydquist & Holm – 38, 39.

Park, Ward & Co – 89.
Parsons, Charles – 30, 31.
Parsons C A & Co – 61.
Parsons Steam Turbine & Co – 31
Power Jets Ltd – 98.
Preston Works – 98.

Railway Magazine – 132.
Ramsey, D M – 34-36.
Reid, Hugh – 34-36.
Riddles, Robert – 10, 97.
Robertson, Brian (Sir) – 9.
Rolls Royce – 26, 27, 89, 132.
Rolls Royce Dart Engine – 142.
Rugby Motive Power Depot – 116.
Rugby Test Centre – 97, 101, 103, 106, 108, 110-118, 139, 155-168.

Scales, B T – 140.
Second World War – 16-20, 24, 40, 152.
Spencer, Bert – 10.
Stamp, Josiah (Sir, later Lord) – 40.
Stanier, William (Sir) – 10-12, 27, 39-42, 53, 126, 153.
Stephenson, George – 16.
Stodola, Aurel – 30, 31.
Sulzer – 83.

Taylor, Taylor & Hobson – 21, 22.
Thomas Ward Ltd – 122.
Turbinia – 32, 33.

Vitry Test Centre – 24.
Vulcan Foundry Works – 104, 110, 122, 132, 135.

Warrington Motive Power Depot – 112, 122.
Whittle, Frank (sir) – 92.
Winterthur- Stein- Sackingen Line – 46.